AN INTRODUCTION TO MOLLUSCA

By
H.S. Bhamrah
Kavita Juneja

Edited by
Balvinder K.

ANMOL PUBLICATIONS PVT. LTD.
NEW DELHI - 110 002 (INDIA)

ANMOL PUBLICATIONS PVT. LTD.
4374/4B, Ansari Road, Daryaganj
New Delhi - 110 002

An Introduction to Mollusca

First Edition, 1991
Reprint, 1992, 1999
Second Revised Edition, 2001

ISBN 81-261-0685-9

PRINTED IN INDIA

Published by J.L. Kumar for Anmol Publications Pvt. Ltd., New Delhi - 110 002 and Printed at Mehra Offset Press, Delhi.

AN INTRODUCTION TO
MOLLUSCA

Contents

PREFACE

An Introduction to Mollusca is the revised and enlarged second edition of the book. It has been designed to approach the morphology, anatomy, physiology and development of selected type in a simple and lucid style. The approach to the discussion of all the types is very simple so as to impact to the students a clear and vivid understanding. According to the scheme of treatment the important animal types of the phylum have been dealt with first, and efforts have been made to present their elaborate and uptodate account. General characters and classification and brief description of other important types of the phylum have also been dealt with complete, authentic and uptodate account. Further separate chapters on topics of significance and general interest pertaining to the phylum have also been added to make the treatment more elaborate.

It has been the constant endeavour of the authors to furnish maximum substance, keeping in view the limitations of size of the volume. Efforts have been made to condense the matter as far as practicable. The book features both a text and a laboratory guide. It is hoped that this book will not only meet the requirement of Indian students but will also be useful as a guideline to the teachers in their teaching.

There can be no claim to originality except in the manner of treatment and much of the information has been obtained from the books and scientific journals available in different libraries.

The authors express their thanks to their friends and colleagues whose constant inspiration have initiated them to bring out this book.

In this edition all the errors of first edition have been rectified. New information is incorporated in the light of recent developments and latest research. Teachers and scholars can send their criticism and suggestions to further improve the contents.

Authors

PREFACE

An Introduction to Mollusca is the revised and enlarged second edition of the book. It has been designed to approach the morphology, anatomy, physiology and development of selected types in a simple and lucid style. The approach to the discussion of all the types is very simple so as to impart to the students a clear and vivid understanding. According to the scheme of treatment the important animal types of the phylum have been dealt with first, and efforts have been made to present their elaborate and uptodate account. General characters and classification and brief description of other important types of the phylum have also been dealt with complete, authentic and uptodate account. Further separate chapters on topics of significance and general interest pertaining to the phylum have also been added to make the treatment more elaborate.

It has been the constant endeavour of the authors to furnish maximum substance, keeping in view the limitations of size of the volume. Efforts have been made to condense the matter as far as practicable. The book features both a text and a laboratory guide. It is hoped that this book will not only meet the requirement of Indian students but will also be useful as a guideline to the teachers in their teaching.

There can be no claim to originality except in the manner of treatment and much of the information has been obtained from the books and scientific journals available in different libraries.

The authors express their thanks to their friends and colleagues whose constant inspiration have induced them to bring out this book.

In this edition all the errors of first edition have been rectified. New information is incorporated in the light of recent developments and latest research. Teachers and scholars can send their criticism and suggestions to further improve the contents.

Authors

1

Mollusca : Characters & Classification

Members of the phylum *Mollusca* are among the most conspicuous invertebrates and include such familiar forms as *clams*, *oysters*, *squids*, *octopods* and *snails.* During the last 18th and 19th centuries, when natural history, occupied the time of many well to do gentlemen, shell collections were as popular as today. Such collections, often containing species gathered from various parts of the world, have contributed considerably to our knowledge of the phylum. In abundance of species, molluscs comprise the largest invertebrate phylum aside from the Arthropoda. Over 1,00,000 living species have been described. In addition, some 35,000 fossil species are known.

The term *Mollusca* (*L. mollis* = soft) mean soft bodied and for the first time coined by *Aristotle* in his famous *Historia Animalium* to describe various kinds of cephalopods and gastropods. The term *Mollusca* was used by *Jonston* (1650) to include cephalopods and barnacles.

Phylum *Mollusca* is divided directly into five classes-Amphineura, Gastropoda, Scaphopoda, Lamellibranchia and Cephalopoda, by *Paseneer* (1892) has remained standard ever since.

Characters

1. Some are aquatic (mostly marine, a few fresh water) and some terrestrial forms.
2. Symmetry bilateral, viscera and shell coiled in Gastropoda and some Cephalopoda.
3. The body is triploblastic made up of three primary germ layers.
4. Body without any appendage and segmentation.
5. Body skin single layered mostly ciliated and with mucous glands.
6. Body usually small and soft and enclosed in an envelop, the *mantle* that secretes shell of various shapes and sizes.
7. Generally the body is divided into 4 parts : the visceral mass carrying the vital organs of the body; the head, anterior most part bearing

mouth, eyes and tentacles but is absent in the *Pelecypoda* and *Scaphopoda*; the foot a muscular mass, the form and function of which varying in the different classes; and lastly the mantle or pallium a thick muscular coat covering the entire animal.

8. Shell, when present, usually univalve, constituting an exoskeleton, internal in some.
9. Coelom is reduced and represented mainly by pericardial cavity, gonadial cavity and kidneys.
10. The alimentary tract is either straight or coiled. Jaws are present in many, especially in gastropods and cephalopods. In most of animals a rasping organ, the *radula* appears in the buccal cavity. The radula bears chitinous teeth to crush food. It is absent in bivalves.
11. Blood vascular system consists of dorsal heart with 1 or 2 auricles and 1 ventricle, and is of open type.
12. Respiration by general body surface or mantle or by one to many gills or a lung in the mantle cavity.
13. Excretion by paired metanephridia (kidneys).
14. Nervous system of paired ganglia, connectives and nerves. Ganglia usually form a circumenteric ring.
15. Sense organs include eyes, statocysts and receptors for touch, smell and taste.
16. Sexes are usually separate (some hermaphrodite a few protandric); gonads 2 or 1 with ducts; fertilization external or internal. Mostly oviparous, few viviparous.
17. Development direct or interrupted with larval stage, the *veliger larva* of mollusca. The asexual reproduction is completely absent.

CLASSIFICATION

Phylum *Mollusca* is divided into six classes. Monoplacophora, Amphineura, Gastropoda, Scaphopoda, Pelecypoda and Cephalopoda.

CLASS-I. MONOPLACOPHORA

1. Abyssal or fossil, from lower Cambrain to Devonian, probably carnivorous.
2. Bilaterally symmetrical and metamerically segmented, but segmentation internal.
3. Shell is dorsal and consisting of a single piece (hence Monoplacophora).
4. Mantle covers the dorsal surface of body.
5. Foot is ventral with a flat creeping sole.

6. Mouth is anterio-median and the anus is posterio-median.
7. Head is without eyes.
8. 5 or 6 pairs of gills are externally and serially arranged.
9. Six pairs of nephridia and two pairs of gonads are present.
10. Stomach contains a *crystelline style* and the animal feeds on radiolarians.
11. Nervous system primitive, lacks ganglia.
12. Sexes are separate.
13. Development is indirect with a *trochophore larva*.

    Example : *Neopilina galatheae*.

CLASS-II. AMPHINEURA

(Gr., *Amphi* = both; *neuron* = nerve)

1. Most primitive molluscs with either elongated and vermiform or dorso-ventrally flattened body.
2. Head is indistinct and without eyes and tentacles.
3. Foot is broad, flat, sole-like and ventral in position.
4. Shell and mantle covering the body from dorsal and lateral sides.
5. 6-60 pairs of gills situated, laterally in the pallial cavity, on either side of foot. They may be reduced to one pair.
6. Mouth and anus are present at the opposite ends of body.
7. Nervous system is primitive and ladder-like with or without ganglia.
8. Heart with one or two auricles.
9. A pair of nephridia connected to the pericardium.
10. Sexes are separate or united.

    Class Amphineura is divided into two orders :

Order-1. Aplacophora (Solenogasters)

1. The body is long, narrow, worm like.
2. The shell is absent.
3. Mantle contains tiny calcareous spicules.
4. Foot is reduced or absent.
5. Heart consists of a single auricle and a ventricle.
6. Gills are absent or reduced to a pair located in the cloacal cavity.
7. Alimentary canal is straight without radula.
8. Nervous system with a distinct brain and ganglia.
9. Sexes united.

    Examples : *Chaetoderma, Neomenia.*

Order-2. Polyplacophora

1. Body elliptical, convex dorsally and flattened ventrally.
2. Shell is dorsal and formed of eight calcareous pieces.
3. Mantle with spicules disposed along the margins.
4. Gills 6 to 8 pairs, present in the pallial groove on the lateral sides of foot.
5. Head is disctinct without eyes and tentacles.
6. The whole ventral surface is occupied by the broad and flat foot.
7. Alimentary canal is coiled with well formed radula.
8. Heart with two auricles.
9. Nervous system without definite ganglia.
10. Sexes separate.

    Examples : *Chiton, Chaetopleura, Cryptochiton.*

CLASS-III. GASTROPODA

(Gr. *Gastros* = stomach; *podo*s = food)

1. Body is asymmetrical with a mantle and shell.
2. The shell is univalved and spirally coiled.
3. Head is distinct bearing tentacles, eyes, and a mouth.
4. Foot is ventral and muscular.
5. Visceral mass usually coiled exhibiting various types of torsions.
6. Anus usually situated anteriorly close to the mouth.
7. Buccal cavity contains an *odontophore* with a radula.
8. Respiration is carried on through the wall of mantle cavity in some, or gills in most forms or lungs.
9. Circulatory system is open type with dorsal heart enclosed in pericardium.
10. Excretory system includes one or two kidneys or nephridia.
11. Nervous system contains distinct cerebral, pleural, pedal, parietal and visceral ganglia.
12. Sexes are separate. Mostly *oviparous*, a few *viviparous*.
13. Development indirect with *veliger* and *trochophore* larval stages.

    Class Gastropoda is divided into three sub-classes. Prosobranchia, Opisthobranchia and Pulmonata

Subclass (a) Prosobranchia (Streptoneura)

1. The visceral nerve commissures are twisted into a figure of 8, hence the name Streptoneura (Gr. *streptos* = curved, *neuron* = nerve). This is due to torsion of visceral hump.

2. The mental cavity containing the pallial complex opens anteriorly.
3. The gills or ctenidia, if present, lie in front of the heart hence Prosobranchia.
4. Head has a single pair of non-retractile tentacles.
5. Calcareous shell with an operculum.
6. Sexes separate. Development through trochophore larva.
7. Exclusively marine.

Subclass Prosobranchia is divided into three orders : Archaeogastropoda, Mesogastropoda and Neogastropoda.

Order-1. Archaeogastropoda (Aspidobranchia)

1. Due to incomplete atrophy of organs of the side, the kidneys, auricles, nephridia and ctenidia are paired and present in mantle cavity.
2. Ctenidia are bipectinate or plume like.
3. Osphradium is poorly developed.
4. The nervous system is little concentrated.
5. Gonad opens into the right kidney serving as gonoduct.
6. Fertilization external.

The order has been distinguished into the following two subroders :

Surborder (i) Docoglossa

1. Shell and visceral mass conical.
2. Ctenidium single or absent or replaced by secondary gills.
3. Radula with few hooked teeth in each row.
4. Auricle single.

Examples : *Patella, Acmaea.*

Suborder (ii) Rhipidoglossa

1. Shell spiral.
2. Ctenidia and auricles are paired.
3. Radula with numerous narrow teeth.

Examples : *Haliotis, Fissurella, Trochus.*

Order-2. Mesogastropoda (Pectinibranchia)

1. These are advanced snails with complete atrophy of the left side so that usually unpaired auricle, nephridium, osphradium and ctenidium occur.
2. Ctenidium monopectinate or comb-like, attached to the mantle throughout its entire length.
3. Edge of the shell opening lacks a siphonal notch or canal.

4. Foot may be operculate.
5. Osphradium pectinate and well differentiated.
6. The nervous system is concentrated.
7. Gonad has its separate duct and opening.
8. Fertilization internal. Development is indirect with free-swimming veliger larva.

   Examples : *Pila, Triton, Valvata, Natica, Vermetus, Cypraea.*

Order-3. Neogastropoda (Stenopoda)

1. Carnivorous species with an eversible proboscis.
2. Osphradium is large and bipectinate.
3. Foot is operculate.
4. One gill with filaments in one row; heart with one auricle; single nephridium are present.
5. Nervous system is lightly concentrated.
6. Shell has a siphonal canal.
7. Embroys are intracapsular. Free swimming veliger suppressed.

   Examples : *Murex, Oliva, Magilus.*

Subclass (b) Opisthobranchia (Euthyneura)

1. Due to detorsion of visceral mass, visceral loop becomes untwisted hence named (Gr. *euthus* = straight; *neuron* = nerve) Euthyneura.
2. Mantle cavity is displaced posteriorly.
3. Head with two pairs of tentacles.
4. Shell and operculum, if present, reduced.
5. Usually one auricle and one nephridium are present.
6. Ctenidia are replaced by secondary gills.
7. Hermaphroditism is universal. Larva is veliger.
8. Exclusively marine.

   Subclass Opisthobranchia is divided into eight orders as follows :

Order-1. Cephalaspidea

1. These are borrowing forms having shield-like head.
2. Shell and mantle cavity moderately developed.
3. Lateral parapodial lobes are prominent.

   Examples : *Bulla, Acteon.*

Order-2. Anaspidea

1. These are crawling or burrowing forms having a pair of *rhinophores* on head.

2. Shell is small and internal.
3. Mantle cavity is reduced on right side.
4. Parapodial lobes are prominent.

Examples : *Aplysia, Akera.*

Order-3. Thecostomata

1. These are shelled pteropods.
2. Shell is spirally coiled or a non-spirally pseudoconch.
3. Mantle cavity is well developed.
4. Parapodial fins are large.

Example : *Limacina, Clio.*

Order-4. Gymnostomata

1. These are naked pteropods and planktonic.
2. Shell and mantle cavity absent.
3. Small ventral parapodial fins are present.

Example : *Pneumoderma.*

Order-5. Notaspidea

1. Shell is external or reduced and internal.
2. Mantle cavity is absent, but a skirt-like projection of mantle covers gill on the right side.

Example : *Pleurobranchus.*

Order-6. Acochlidiacea

1. Small without shell and gill.
2. Naked visceral mass projecting behind the foot and covered with spiracles.

Examples : *Acochlidium.*

Order-7. Sacoglossa

1. These are herbivorous, with a modified radula and suctorial pharynx.
2. They are shelled or naked slug-like, with a gill.

Examples : *Elysia, Oxynoe.*

Order-8. Nudibranchia (Acoela)

1. They are naked bilaterally symmetrical sea-slugs without shell, mantle cavity, ctenidia and osphradium.
2. They have secondary gills around anus or surface outgrowths or cerata.

Examples : *Doris, Aeolis, Tritonia.*

Subclass (c) Pulmonata

1. The ctenidia are absent and replaced by pulmonary sac or lung.
2. Shell is simple spiral or vestigial or absent and operculum never occur.
3. One or two pairs of tentacles and one pair of eyes are present.
4. The torsion takes place. The nervous system is secondary symmetrical due to shortening of connectives and concentration of ganglia into a circumoesophageal ganglionic complex.
5. The heart with a single auricle lying anterior to the ventricle.
6. The animals are bisexual. Gonad single.
7. Development direct without larval stage.

The subclass is divided into two orders :

Order-1. Basommatophora

1. One pair of tentacles with the eyes at the base.
2. Shell is delicate with a large aperture.
3. Some forms with secondary gills.
4. Male and female gonopores usually separate.

Examples : *Limnaea, Planorbis, Physa.*

Order-2. Stylommatophora

1. Two pairs of retractile tentacles. Eyes are lodged on the tips of posterior pair of tentacles.
2. Shell is internal, reduced or absent.
3. Mostly terrestrial.
4. Male and female gonopores usually united.

Examples : *Helix, Limax, Arion.*

CLASS-IV. PELECYPODA

1. The Pelecypoda or Bivalvia are marine or fresh-water, bilaterally symmetrical molluscs.
2. Mantle consists of paired right and left lobes, which secrete a bivalved shell.
3. Head, eyes, tentacles and radula are absent.
4. The muscular foot is usually wedge-shaped or tongue-shaped, which is adapted for burrowing.
5. Presence of two gills, one on either side of mantle cavity, the function of which is to produce respiratory and food carrying currents of water.
6. Pharynx is absent.
7. Coelom is reduced to a dorsally placed pericardium.
8. Alimentary canal is coiled with a large paired digestive glands.

9. Heart is continued with pericardium and comprises two auricles and one ventricle.
10. The renal organs are a pair of coelomic kidneys, which lead into pericardium at one end and to outside at other.
11. The nervous system includes four pairs of ganglia, viz., cerebral, pleural, pedal and visceral.
12. Sense organs are statocyst and osphradium.
13. Sexes may be separate or united. Development includes metamorphosis, followed by trochophore stage.

Class - Pelecypoda is divided into following orders :

Order-1. Protobranchia

1. Presence of paired plume-like gills, each bears two rows of filaments.
2. The foot is flattened.
3. Presence of anterior and posterior adductor muscles.
4. Sexes separate, gonads lead into kidneys.

Examples : *Nucula, Yoldia.*

Order-2. Filibranchiata

1. Single pair of plate-like gills formed of distinct V-shaped filaments.
2. Inter-filamentar junctions are either absent or formed by groups of inter-locking cilia.
3. Inter-lamellar junctions are either absent or non-vascular.
4. Two adductor muscles are present, anterior may be reduced or absent.
5. Foot small or poorly developed.

Example : *Mytilus, Arca.*

Order-3. Pseudolamellibranchiata

1. The gills are plaited so as to present vertical folds.
2. Inter-filamentar junctions may be ciliary or vascular. Interlamellar junctions may be vascular or non-vascular.
3. Shell valves are frequently unequeal and single large adductor muscle present.
4. Foot is slightly developed or rudimentary.
5. Genital organs open into kidneys or near them.

Examples : *Pecten, Ostrea, Pinna, Lima.*

Order-4. Eulamellibranchiata

1. Gills are basket like, the gill-filaments are attached with vascular inter-filamentar and inter-lamellar junctions.
2. Gills may be smooth or plaited.

3. Two equal adductor muscles.
4. Foot is well developed.
5. Siphon of small or large sized present.
6. Gonads open out by separate openings. They do not open into kidneys.

   Examples : *Unio, Teredo, Anodonta, Venus, Cardium, Mya.*

Order-5. Septibranchiata

1. Gills replaced by horizontal muscular and perforated septa dividing the mantle cavity into upper and lower chambers.
2. Foot long and slender, byssus reduced or absent.
3. Adductor muscles two.
4. True ctenidia are absent.

   Examples : *Cuspidaria, Poromya.*

CLASS-V. SCAPHOPODA

1. They are aberrant marine animals comprising only of three genera.
2. Body is elongated worm-like, enclosed in a tusk-like shell open at both the ends.
3. Eyes, tentacles and gills are absent.
4. Mantle tubular, completely enclosing the body.
5. Mouth surrounded by lobular processes or outgrowths.
6. Foot is reduced used for digging.
7. Heart is rudimentary.
8. Sexes are separate.
9. Torsion takes place in some forms.

   Scaphopoda has two families :

Family (i) Dentaliidae

1. Foot is trilobed having two epipodial lobes and one hypopodial lobe.

   Examples : *Dentalium, Antalis.*

Family (ii) Siphonodentalidae

1. Foot is elongated and capable of expansion into a terminal disc.

   Examples : *Pulsellum, Siphonodentalium.*

CLASS-VI. CEPHALOPODA

1. Exclusively marine.
2. Shell is usually internal and reduced, absent shell embedded in mantle.
3. Body with head and trunk. Head bears eyes and mouth.
4. Trunk consists of symmetrical and uncoiled visceral mass.

5. Mantle encloses posteriorly and ventrally a large mantle cavity.
6. Foot altered into a series of sucker bearing arms or tentacles encircling the mouth.
7. Mouth bears jaws and radula.
8. Two or four pairs of bipectinate gills.
9. Circulatory system is closed, heart with two or four auricles.
10. Excretory system comprises two or four pairs of nephridia.
11. Nervous system is highly developed and the principle ganglia are concentrated around the oesophagus.
12. Sexes are separate.
13. Development meroblastic without metamorphosis.
14. Most of the cephalopods have an *ink-gland* with a duct opening into the rectum.

Cephalopoda is divided into three subclasses :

Subclass (a) Dibranchiata

1. The foot modified into a circlet of eight to ten arms bearing suckers, around the mouth.
2. The funnel forms a complete tube.
3. The shell is generally internal.
4. Presence of paired gills and kidneys.
5. Presence of characteristic ink-gland and duct.

Dibranchiata is divided into two orders.

Order-1. Octopoda

1. Body rounded or oval and with no lateral fins.
2. Arms are eight, long arms or tentacles absent.
3. Suckers on the arms sessile and without horny rim.
4. Shell and cnidamental glands absent.

Examples : *Octopus* (Devil-fish), *Agronauta.*

Order-2. Decapoda

1. Body elongated and with no lateral fins.
2. Arms ten, eight short and two long and known as tentacles.
3. Suckers on the arms are sessile and without horny rims.
4. Shell is internal and well developed.

Examples : *Sepia, Loligo.*

Subclass (b) Nautiloidea or Tetrabranchiata

1. The foot is divided into lobes bearing numerous tentacles.

2. The funnel is incomplete.
3. Shell is external which is divided into several compartments.
4. Kidneys and gills are four.
5. Devoid of ink gland.
6. Pericardium communicates to the exterior directly.
7. Eyes are simple without crystelline lens.
8. Chromatophore and salivary glands are absent.
   Example : *Nautilus.*

Subclass (c) Ammonoidea

1. All are extinct.
2. Shell variously modified in the form of straight to spiral or turreted.
3. Siphon external, siphuncle simple and marginal.
4. Two pairs of ctenidia.
   Example : *Ammonities* (Extinct).

CHAETODERMA

Phylum	-	Mollusca
Class	-	Amphineura
Subclass	-	Aplacophora
Genus	-	*Chaetoderma*

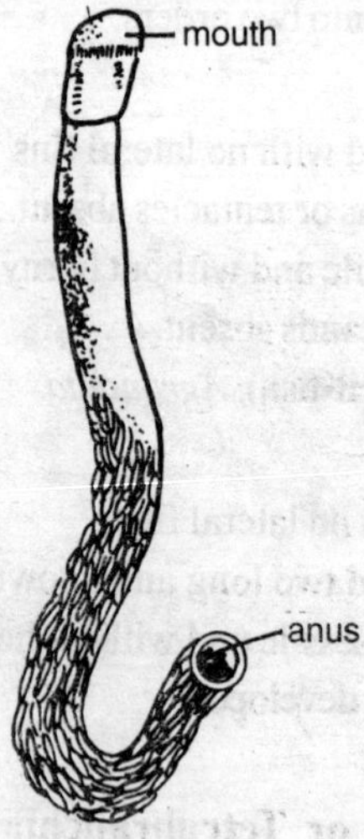

Fig. 1.1. *Chaetoderma.*

1. It is a most primitive mollusc. The body is worm-like cylindrical. It is found around North Atlantic.
2. The head is separated off from the body by a constriction.
3. Mantle covers the body completely.
4. Shell is absent but the body is covered by numerous calcareous spicules.
5. Rudimentary foot is present.
6. Gills reduced to a pair which are situated in the cloacal cavity.
7. Sexes are separated and each has one fused gonad.
8. A cavity is present at the posterior end of the body into which opens the anus.

PATELLA

Phylum	-	Mollusca
Class	-	Gastropoda
Subclass	-	Prosobranchia
Order	-	Aspidobranchia
Suborder	-	Docoglossa
Genus	-	*Patella*

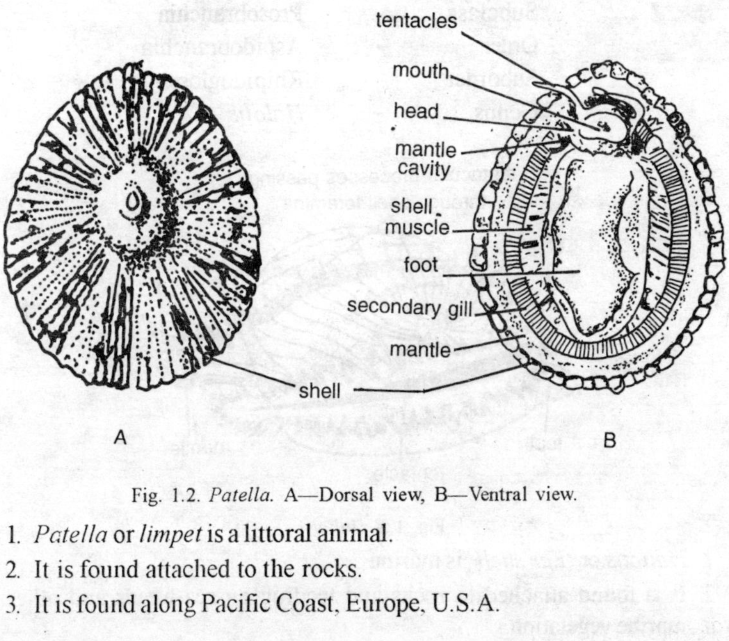

Fig. 1.2. *Patella.* A—Dorsal view, B—Ventral view.

1. *Patella* or *limpet* is a littoral animal.
2. It is found attached to the rocks.
3. It is found along Pacific Coast, Europe, U.S.A.

4. Body is oval with convex dorsal surface.
5. Shell is oval and core-like, with a conical projection from its dorsal surface.
6. Operculum is absent.
7. The head is laterally produced into a pair of small, stout, tactile tentacles.
8. Eyes are simple present at the base of tentacles.
9. Foot is ventral and sole-like for creeping.
10. True mantle is restricted to the anterior region and secondary mantle cavity is developed between foot and mantle.
11. Radula composed of very few, strong hooked teeth in each row.
12. All around the foot lies a secondary pallial cavity containing a series of secondary branchiae.
13. Heart with single auricle.
14. Herbivorous, feeding upon algae, sea-weeds and diatoms.
15. It is eaten by poor classes in country like France, Italy and Ireland.

HALIOTIS

Phylum	-	Mollusca
Class	-	Gastropoda
Subclass	-	Prosobranchia
Order	-	Aspidobranchia
Suborder	-	Rhipidoglossa
Genus	-	*Halotis*

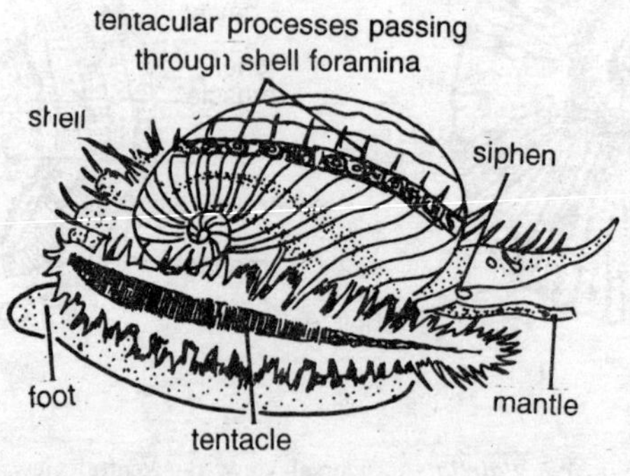

Fig. 1.3. *Haliotis.*

1. *Haliotis* or *'Ear shell'* is marine
2. It is found attached to rocks and feeding on sea-weeds and other marine vegetation.

3. The animals are very common at and below low tide level.
4. *Haliotis* or *Abalone* has ear-shaped shell with small flattened spire.
5. The aperture is very large and without operculum.
6. The shell is perforated by a series of marginal slits through which project the tentacular processes of the mantle.
7. The foot is large and with epipoidia and the animal can move 2 to 4 meters per minute but clings very powerfully to the boulders.
8. Eyes are borne on stalks at the outer bases of the tentacles.
9. The mantle cavity is large, has bipectinate ctenidia or gills, the right being smaller.
10. The shell of the animal is of great economic importance, as beautiful pearly buttons are manufactured from the shell of *Haliotis* in Europe and U.S.A.
11. The animal is also eaten in countries like China, Japan and many cities along the Pacific Coast.

FISSURELLA

The systematic position is the same as that of *Halotis.*

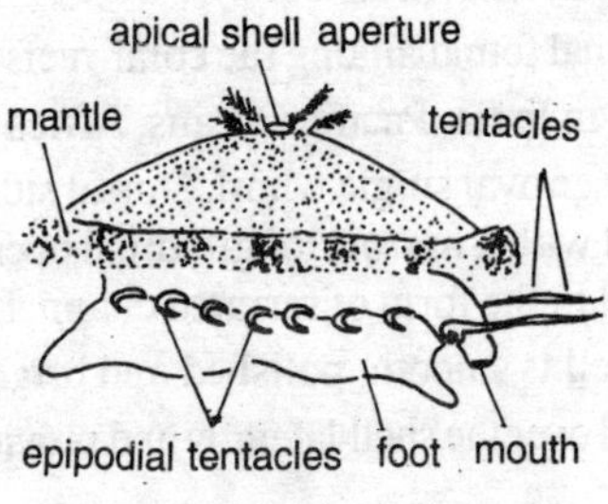

Fig. 1.4. *Fissurella.*

1. It is commonly known as *'Key-hole limpet'*.
2. It is a marine gastropod found attached to the rocks below low-tide mark.
3. The shell is oval, compressed, conical with an aperture at the apex which opens into the branchial chamber. The operculum is absent.
4. Ventral foot is large, sole-like and is used as sucker for clinging to rocks.
5. Foot bears on either side an epipodial ridge bearing a row of tentacles or cirri.
6. Head bears two small, stout and sensory tentacles.

CYPRAEA

Phylum	-	Mollusca
Class	-	Gastropoda
Subclass	-	Prosobranchia
Order	-	Mesogastropoda
Genus	-	*Cypraea*

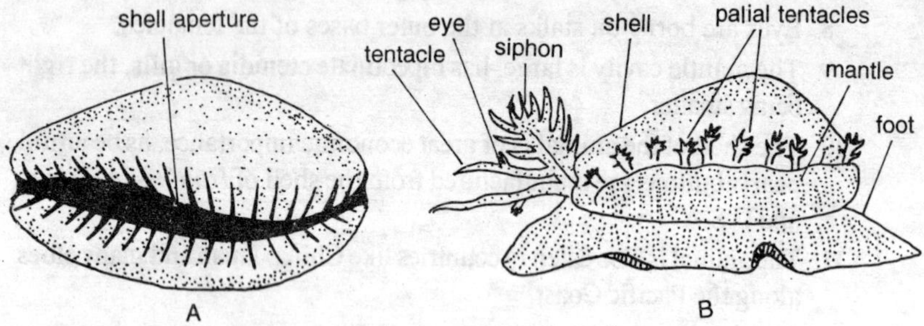

Fig. 1.5. *Cypraea.* A—External shell; B—Entire animal.

1. *Cyprea* is commonly known as *'cowry'*.
2. It is a marine animal found among the coral-reefs and rocky ground.
3. It is mostly found in India, Pacific Oceans, Africa and U.S.A.
4. The shell has a large oval single whorl. Dorsal side is convex and the ventral side is flat with a narrow longitudinal opening extending the length of the shell in the form of serrated channel.
5. The surface of shell is smooth, polished and brightly coloured.
6. Mantle is reflected over the shell laterally and provided with tentacular filaments.
7. Operculum is absent.
8. In entire expanded animal, a pair of eyes, tentacles, pallial tentacles, foot and siphon are easily seen.
9. The shells of *Cypraea* are used for ornaments, the *Cypraea moneta* is used as money.

MUREX

Phylum	-	Mollusca
Class	-	Gastropoda
Subclass	-	Prosobranchia
Order	-	Pectinibranchia
Genus	-	*Murex*

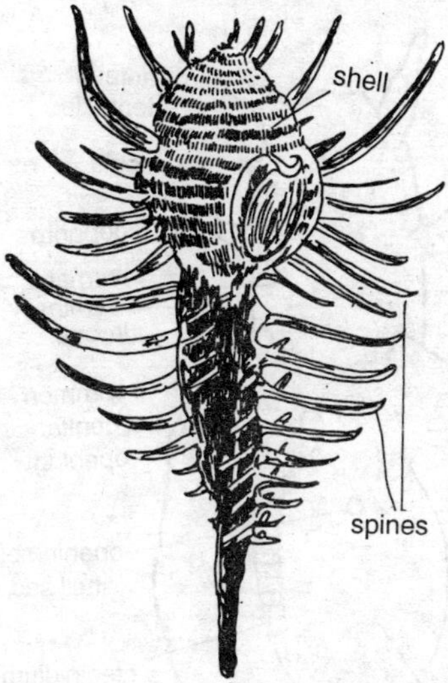

Fig. 1.6. *Murex.*

1. It is a marine animal found between low tide-marks on stones and sea weeds.
2. It is founded on the Syrian Coast, Italy, India, West Indies and U.S.A.
3. Shell is spirally coiled and with three or more rows of beautiful spines of different lengths.
4. It is carnivorous and highly predator feeding on living animals which it grasps by means of its foot. It is destructive to oyster bed.
5. A very prominent bipectinate osphradium perist.
6. Head is prolonged into a long proboscis which can be retracted within the proboscis-sheath.
7. *Adrectal glands* present which secrete a dye the Jyrain purple (Yellow in colour).
8. Sexes are separate.

APLYSIA

Phylum	-	Mollusca
Class	-	Gastropoda
Subclass	-	Opisthobranchia
Order	-	Anaspidea
Genus	-	*Aplysia*

1. It is a marine animal found among sea-weeds.

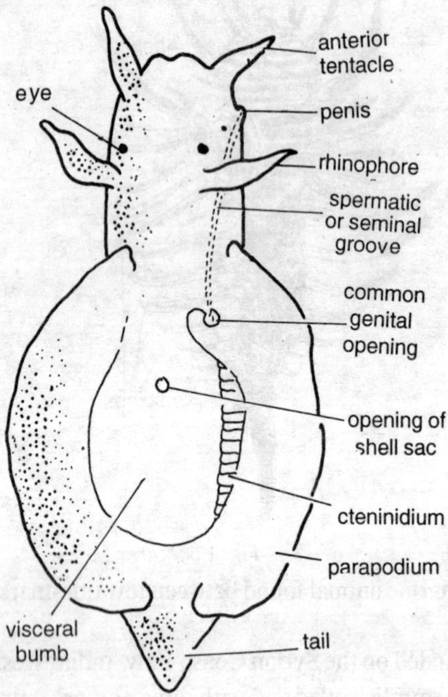

Fig. 1.7. *Aplysia*

2. It is found in India, Asia, West Indies and on the Florida Coast.
3. It is commonly called *'sea hare'*.
4. Body is soft, lumpy with a visceral hump.
5. Head with two pairs of tentacles. The anterior pair is large and ear-like, the posterior pair is olfactory and bear eyes at their bases.
6. Shell thin, reduced and transparent, enclosed by mantle.
7. The foot is muscular and elongated pointed posteriorly with lateral out growths, the *parapodia*.
8. Mantle possesses unicellular ink-glands secrete purple ink, used for defence.
9. Hermaphrodite. There is a single genital aperture and a single duct for the sperms and ova but a seminal groove runs from the aperture to the head and reciprocal fertilization is impossible.
10. Nervous system is well developed with perfectly symmetrical visceral loop.

EOLIS

Phylum	-	Mollusca
Class	-	Gastropoda
Subclass	-	Opisthobranchia
Order	-	Nudibranchia
Genus	-	*Eolis*

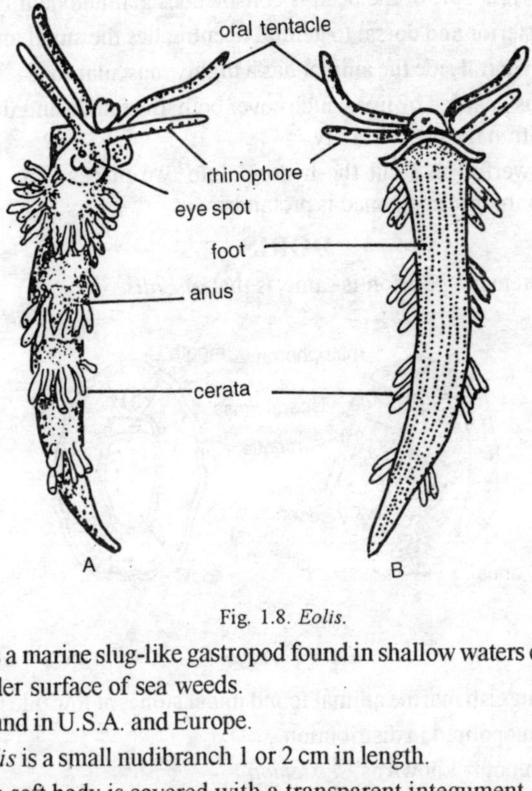

Fig. 1.8. *Eolis.*

1. It is a marine slug-like gastropod found in shallow waters crawling on under surface of sea weeds.
2. Found in U.S.A. and Europe.
3. *Eolis* is a small nudibranch 1 or 2 cm in length.
4. The soft body is covered with a transparent integument.
5. The shell, mantle and true gills are absent.
6. The *cerata* are the respiratory organs.
7. There are cylindrical extensions of the integument, containing hollow extensions of the liver and nematocysts derived from ingested hydroids.
8. The cerata have characteristic arrangement, may occur in many transverse or in few longitudinal rows.

9. The head is small, bears tentacles, usually two pairs, one may be retractile.
10. The anterior or oral pair of tentacles is longer.
11. The posterior or dorsal pair, known as *rhinophores* bear simple sense organs which taste the surrounding water.
12. At the base of each rhinophore is an *eyespot*.
13. The mouth is prominent and can be easily seen from the ventral view.
14. On the right side of the head is conspicuous genital opening.
15. Just posterior and dorsal to genital opening lies the small anus.
16. On the ventral side the animal has a highly muscular foot.
17. Mouth is guarded by lips which cover both, the radula and the horny jaws within the buccal cavity.
18. The powerful jaws cut the hydroid into two pieces. The *Eolis* is hermaphroditic and gonad is protandric.

DORIS

The systematic position is same as that of *Eolis.*

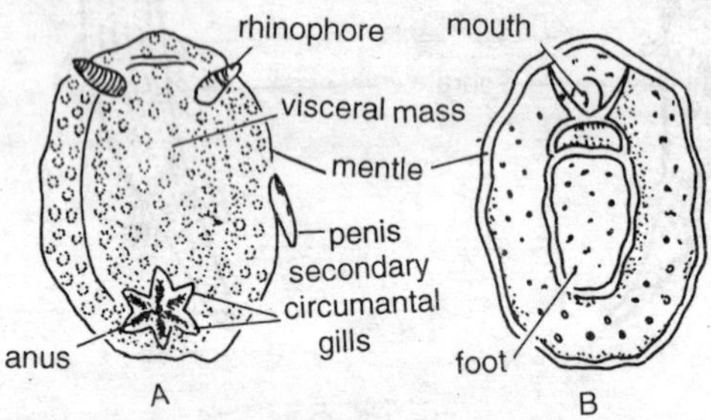

Fig. 1.9. *Doris.*

1. It is a sluggish marine animal found under stones at low tide marks.
2. It is cosmopolitan in distribution.
3. It is commonly known as *'sea lemon'*.
4. The body is oval and depressed, concave above and flat below.
5. Head is indistnict, but there is a pair of short olfactory tentacles or rhinophores towards the anterior end.
6. Mantle is usually pigmented and contains calcareous spicules or dorsal tubercles.
7. Mouth is present on ventral side. Anus lies mid-posteriorly and is surrounded by a circlet of feathered retractile secondary branchiae.

8. The broad foot is on ventral side.
9. Hermaphrodite, gonopore is asymmetrically placed on the right side of the body.
10. *Doris* has undergone detorsion, therefore, the shell, mantle cavity and primary gills are absent.

LYMNAEA

Phylum	-	Mollusca
Class	-	Gastropoda
Subclass	-	Pulmonata
Order	-	Basommatophora
Genus	-	*Lymnaea*

1. It is commonly known as *pond-snail.*
2. It is a fresh water snail found in ponds, lakes etc.
3. It is found from Northern states from Atlantic to Pacific, Europe, California, New England to Kansas and entire Pacific Coast.
4. Shell is thin, horny and fragila with a sharply elongated spire.

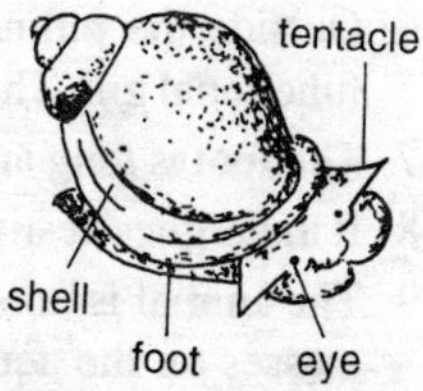

Fig. 1.10. *Lymnaea*

5. Body whorl is large with an aperture, while the *penultimate whorl* and apex are smaller.
6. Operculum is present.
7. Tentacles are broad and flat bearing eyes at the base.
8. Foot is rounded behind.
9. Aerial respiration by pulmonary sac.
10. They can survive for sevaral weeks in ice.
11. Species of *Lymnaea* serves as intermediate host for the larvae of *Fasciola hepatica.* Sporocyst and redia stages of *Fasciola* are found in *Lymnaea truncatula.*

PLANORBIS

The systematic position of *Planorbis* is same as that of *Lymnaea.*

1. It is a fresh water animal found in freshwater ponds attached to submerged plants.
2. It is commonly found in India and Europe.
3. The shell is same what black, stout and spirally coiled.
4. Shell is *sinistral* and *discoidal* having a depressed or flattened spire.

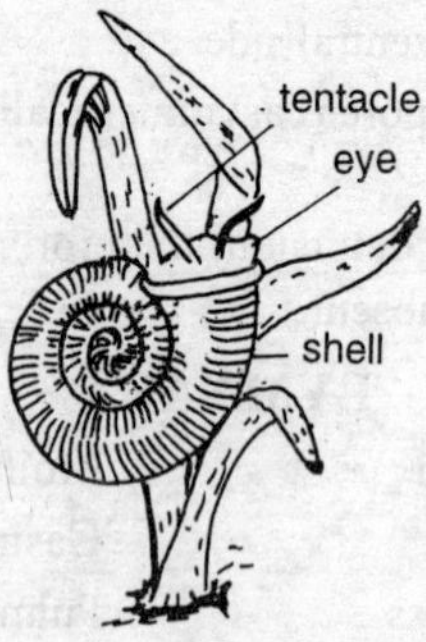

Fig. 1.11. *Planorbis.*

5. Head protude through the shell opening. It bears filiform tentacles, and a pair of eyes.
6. Outside the pulmonary cavity, the mantle lobe is modified into a functional gill. The plasma contains haemoglobin.
7. The foot is long and muscular.
8. It undergoes aestivation during summer.
9. The animal is unisexual. The eggs are laid in the form of gelatinous masses on the aquatic plants or stones.
10. It serves as intermediate host for liver-fluke and schistosomes.

LIMAX

Phylum	-	Mollusca
Class	-	Gastropoda
Subclass	-	Pulmonata
Order	-	Stylommatophora
Genus	-	*Limax.*

1. It is a terrestrial gastropod, found in gardens over damp soil.
2. It is cosmopolitan in distribution.
3. It is commonly known as *'gray slug.'*
4. Body is elongated and tapering behind and is divisible into head, foot and visceral hump.
5. Two pairs of retractile tentacles are present on head. Eyes are present at the tips or posterior tentacles.
6. The mouth is on the ventral side of the anterior end and is bounded by a pair of lateral lips.
7. Mantle forms a shield-like area in the dorsal side of the anterior part of the body. The mantle cavity acts as a lung for the breathing; pneumostome is present on right side.

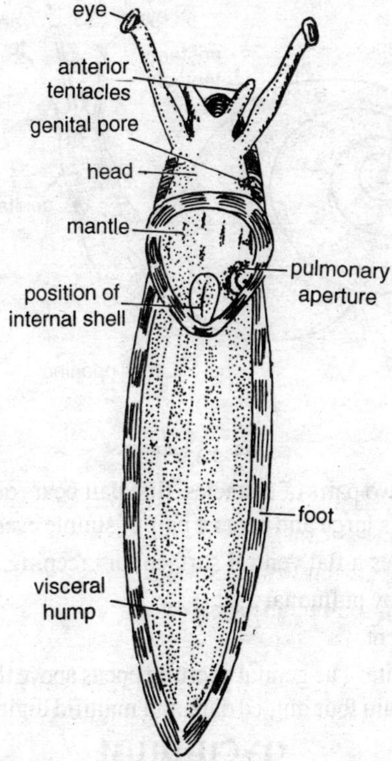

Fig. 1.12. *Limax.*

8. Shell is internal, thin and rudimentry, lies embedded in mantle.
9. Foot is on ventral side.
10. A pedal gland just behind the head for the mucous secretion.
11. The animal is bisexual.
12. Nocturnal and herbivorous.

HELIX

The systematic position is same as that of *Limax*.

1. It is a terrestrial animal found in damp places, under bark and fallen leaves.
2. It is found in Palearctic region.
3. It is commonly called *'garden snail'*.
4. The shell is thin, external and *heliciform* with a low conical spire, measuring 4 cm. in length.

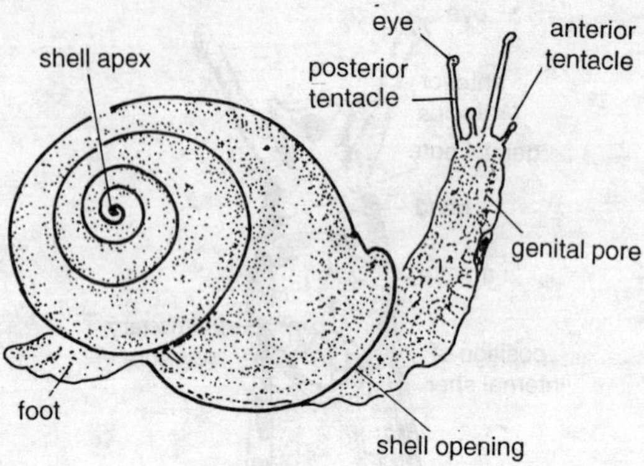

Fig. 1.13. *Helix.*

5. Head bears two pairs of tentacles, first pair bears organs of small and second pair is large and bears a pair of simple eyes.
6. Foot possesses a flat ventral surface for creeping.
7. Respiration by pulmonary sac.
8. Gills are absent.
9. Hermaphrodite. The genital aperture opens above the right lateral lip. Gonads contain four angled dark and multifid digitiform glands.

ONCHIDIUM

The systematic position is the same as that of *Limax.*

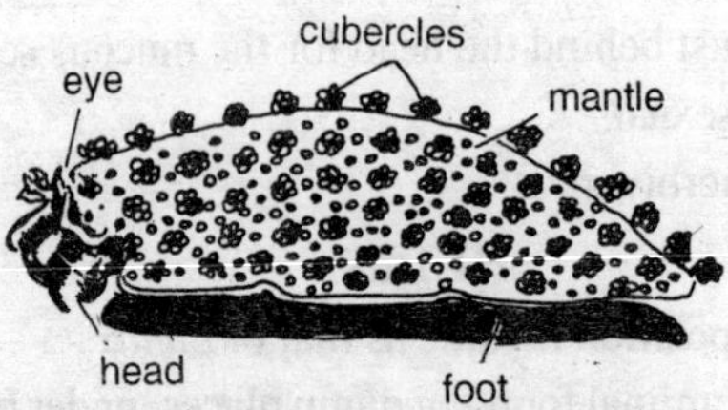

Fig. 1.14. *Onchidium.*

1. It is a slug-like marine animal lives on the sea coasts in brackish marshes.
2. Body is elongated or oval with a convex dorsal surface.

3. The entire dorsal surface is covered by mantle. The mantle is thick and bears numerous wart-like tubercles.
4. Eyes are present on eye-stalks but the tentacles are absent. Accessory eyes are also present on the tubercles of back. These back or dorsal eyes prevent the animal from its enemies.
5. Mantle cavity and ctenidium are absent.
6. A pulmonary chamber or lung is present in mantle. The lung opens through a small contractile aperture called *pneumostome*.
7. Foot is broad with a creeping sole on ventral side.
8. Hermaphrodite. The female genital pore lies posteriorly near the anus while the male genital pore is present at the base of right eye-stalk.

MYTILUS

Phylum	-	Mollusca
Class	-	Pelecypoda
Order	-	Filibranchiata
Genus	-	*Mytilus*

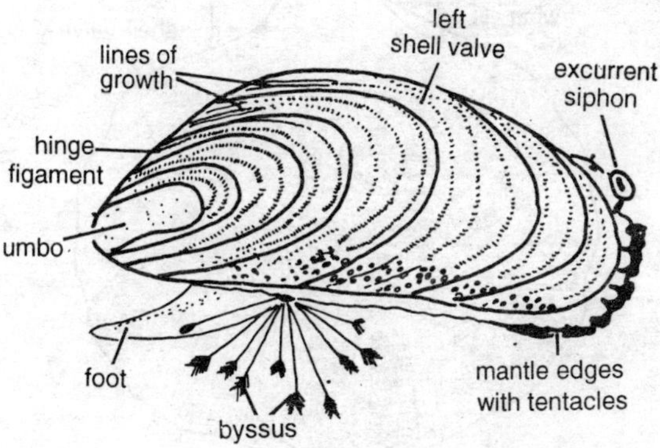

Fig. 1.15. *Mytilus.*

1. It is a marine, sedentary mollusc found at a depth of 2 or 3 fathoms in low tide-mark, attached to rocks or wooden structures by its byssus threads.
2. It is commonly called as *'sea mussel'* and cosmopolitan in distribution.
3. The shell is elongated, equivalent with umbo at or near anterior end.
4. Hinge toothless but may bear cerculations.
5. The foot is tongue-like and is modified into a bunch of *byssal threads*

coming out from between the two shell valves, which are main anchoring organs.

6. The anterior adductor muscle is weaker where as posterior is strongly developed.
7. A pair of gills is present. Each gill is provided with gill- filaments. The gills are lamelliform.
8. The eyes are found anterior to the inner gill-lamella..
9. Sexes are separate. Gonads extend into the mantle.
10. It is used as food in Europe.

PECTEN

Phylum	-	Mollusca
Class	-	Pelecypoda
Order	-	Pseudolamellibranchiata
Genus	-	*Pecten*

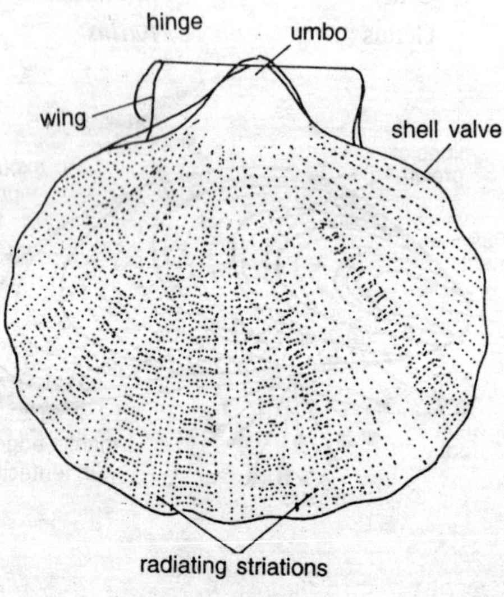

Fig. 1.16. *Pecten.* Dorsal view.

1. It is a free swimming marine mollusc lives on the sea bottom of 10 fathoms deep.
2. It is found in India and U.S.A.
3. It is commonly called as *scallops*.
4. Two shells are unequal. The right shell being larger and more convex and the animal rests on this valve.

5. There is a single large adductor muscle. This is divided into two parts and the larger of these served for rapid contraction which cause swimming movements.
6. Foot is very much reduced but in larva very well developed and used for locomotion.
7. Two large and crescentic gills are present.
8. Mantle is *tentaculiferous* and it encloses viscera.
9. Large number of stalked eyes are present at regular intervals along the edges of the mantle.
10. These are hermaphrodite.
11. The ovary has pink colour when eggs are ripe.
12. The testis lies behind and it is cream in colour.

OSTREA

The systematic position is same as that of *Pecten*.

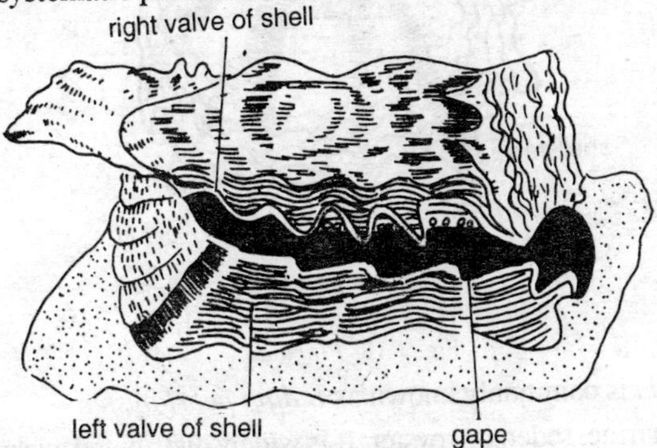

Fig. 1.17. *Ostrea*.

1. It is a sedentary bivalve attached to rocks or other shells and found in shallow and brackish water.
2. It is found in Atlantic and Pacific Coasts, India, Gulf of Mexico to Massachusetts.
3. It is commonly called '*edible oyster*'.
4. The shell valves are unequal the left is always larger and permanently attached to the rocks by byssus.
5. Only one adductor muscle i.e. posterior one. It is divided into two parts one is straited and other is non-straited.
6. The foot is absent.
7. Heart with two fused auricles.
8. A pair of curved gills are present.

9. It is peculiar in the sense as the individuals function alternately as males and females.
10. Spawning generally takes place at full moon. The eggs develop into veliger larva which metamorphoses into adult.

PINCTADA

Systematic position is the same as that of *Pecten.*

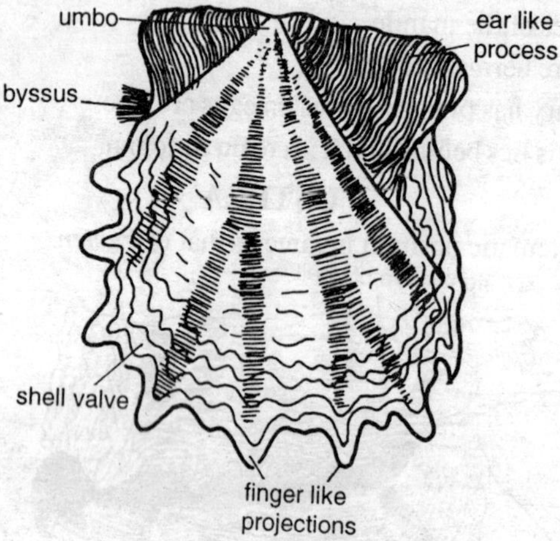

Fig. 1.18. *Pinctada.*

1. *Pinctada* is commonly known as *Indian pearl oyster.*
2. It is a marine, sedentary oyster. It is widely distributed except in cold zones. It is very common in Gulf of Kutch, Tropical and Temperate seas, Pacific and Atlantic coasts.
3. Shell valves are unequal and the hinge line is straight produced at each end into a wing-like process.
4. The left shell is large, convex and permanently attached to rocks by a strong byssal thread.
5. Right shell is smaller and thinner and covers viscera.
6. There is a single adductor muscle like that of *Pecten.*
7. The right and left mentle lobes are free.
8. The sexes are separate but probably alternate in the same individual in successive sessions.
9. The shell secretes pearl whenever any organism enters the shell and causes irritation.

TEREDO

Phylum	-	Mollusca
Class	-	Pelecypoda
Order	-	Eulamellibranchiata
Genus	-	*Teredo*

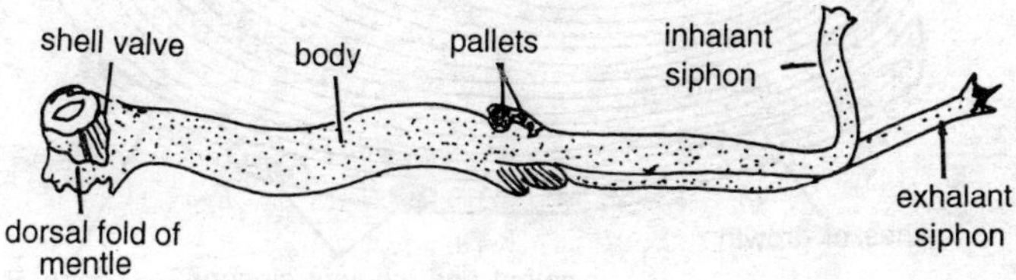

Fig. 1.19. *Teredo.*

1. Marine pelecypod which burrows in wood of the boats, ships and other submerged wood.
2. It is found in India, Europe, Cape Cod to Florida and Massachusetts Bay to Florida.
3. It is commonly called '*shipworm*'.
4. *Teredo* has extremely elongated body and worm-like appearance.
5. Foot is reduced.
6. Mantle tubular and opens anteriorly, shell is reduced and covers only the anterior part of the body.
7. The mantle cavity and siphons are long and the latter unites to a large extent and provided with calcareous pellets for protection.
8. A constant current into and out of the mantle cavity is maintained by ciliary action.
9. The saw dust is swallowed by the animal and is digested by cellulose digesting enzymes.

MYA

The systematic position is same as that of *Teredo*.

1. It is commonly known as *'mud-clam'*.
2. It a marine and burrowing animal found between lives and in shallow water in mud flats and under stones.
3. It is found in Arctic Ocean to Cape Hatteras, San Francisco and Europe.
4. It resembles *Unio*. Body is covered with equall shell valves with hinge ligament.
5. The periostracum is wrinkled.

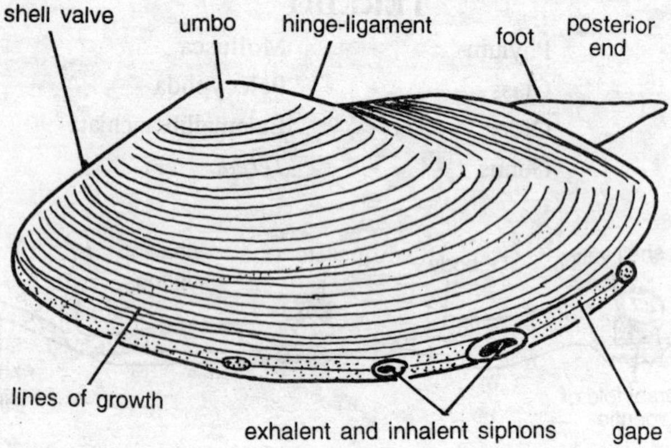

Fig. 1.20. *Mya.*

6. On each shell there is a raised white area called *umbo*.
7. Foot is feebly developed and the byssus thread is absent in adult.
8. Inhalent and exhalent siphons are well developed. The siphons are long and covered with chitinous plates and incompletely retractile.
9. Eyes, tentacles and radula are absent. The adductor muscles are *isomyarian*.
10. *Mya* is the food of walrus.

VENUS

The systematic position is the same as that of *Teredo*.

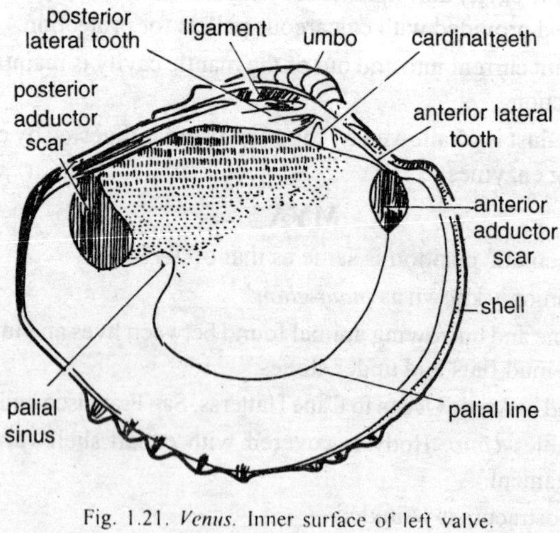

Fig. 1.21. *Venus.* Inner surface of left valve.

1. *Venus* is commonly known as *'hard-shelled clam'* or *'little neck clam'*.
2. It is a burrowing form. It occurs in muddy and sandy flat near the lower position of inter-tidal zone of Atlantic Coast. It is found in India, Europe and U.S.A.
3. The shell comprises two symmetrical halves. Dorsally two shells are joined by hinge ligament.
4. Shell is heart-shaped. The anterior end is short, posterior end rounded and bears lines of growth.
5. Umbo directed forwards, shell-surface dirty white with prominent sharp concentric ridges. The inner shell surface is smooth, dull white with violet or purple lower margin with granules.
6. Siphons are short, unequal and more or less separated.
7. The mantle margins are fringed.
8. Lunule is a heart-shaped configuration which lies anterior and ventral to beak.
9. Spawning in rainy seasons. Millions of eggs and sperms are produced. Young clam lies dormant for two years.
10. Extensively used as food. It is also used in manufacture of wampum, the violet portion forming the most valuable piece.

ENSIS

The systematic position is same as that of *Teredo*.

1. It is commonly known as *'razor-clam'* or *'razor shell'*.
2. It is a burrowing animal found in the sand about the low tide mark. It can also swim backwards by jerks of foot.
3. It is found in India and U.S.A.
4. The animal is covered by a thin shell. The shell is made up of two valves which are slightly bent in the middle with concave dorsal and convex ventral sides.
5. The shell measures 15 cm. long and 25 mm. high.
6. The right valve has two hinge teeth and the left valve has three hinge teeth.
7. The mantle lobes are united ventrally to form an elongated tubular mantle cavity.
8. Siphons are more or less small and united about half of their length.
9. A pair of long and narrow gills is present in the mentle cavity.
10. Foot is more or less cylindrical and muscular organ produced from the anterior end, helps in burrowing.
11. Muscles are *isomyarian*.

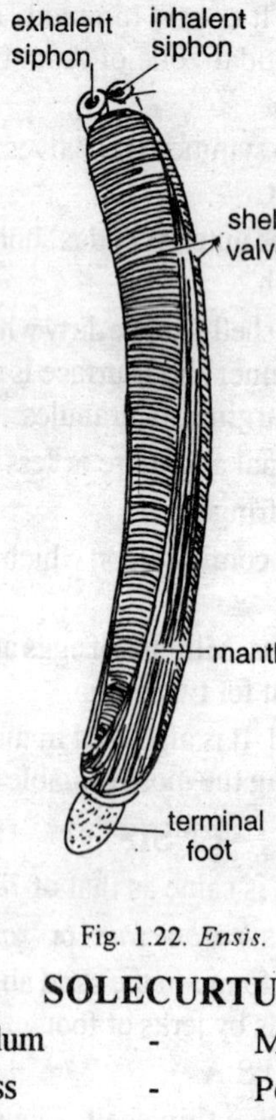

Fig. 1.22. *Ensis.*

SOLECURTUS

Phylum	-	Mollusca
Class	-	Pelecypoda
Order	-	Eulamellibranchiata
Genus	-	*Solecurtus*

1. It is a burrowing pelecypod found in India and U.S.A.
2. It is commonly called '*razor fish*' or '*razor shell*'.
3. Bivalve shell, mantle divides into right and left lobes.
4. Body is laterally compressed and bilaterally symmetrical.
5. A pair of basket-shaped gills consisting of filament, which are united by inter-filamentar and inter-lamellar junctions are present.
6. Two equal-sized adductor muscles are developed.
7. Presence of two elongated siphonal tubes which are united with each other.
8. The shell has the appearance of long scabbard.

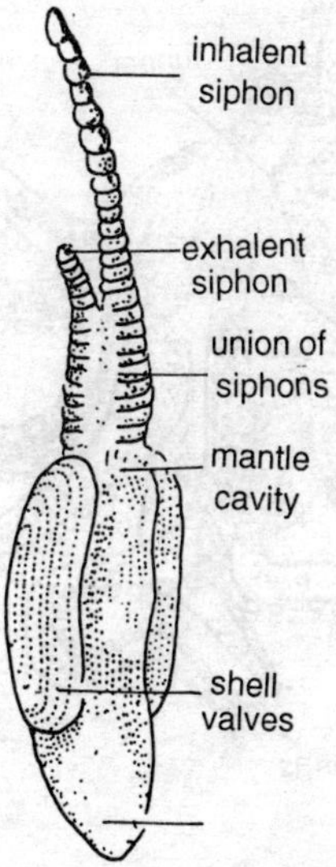

Fig. 1.23. *Solenocurtus*. (Razor-fish).

9. The foot is long, cylindrical and enormously developed which is capable of penetrating into the loose sand.

OCTOPUS

Phylum	-	Mollusca
Class	-	Cephalopoda
Subclass	-	Dibranchiata
Order	-	Octopoda
Genus	-	*Octopus*

1. It is marine and cosmopolitan in distribution generally found on the Atlantic and Pacific Coasts.
2. It is commonly called *'devil fish'*.
3. Body surrounded by mantle containing chromatophores. They can change their colour according to surrounding medium.
4. Foot modified into eight equal arms surrounding the head and siphon.
5. The arms are beset with suckers which are sessile and without horny rims.
6. Body is covered with tubercles, short, round aborally and without fins.
7. External shell is absent and body with vestiges of an internal shell.

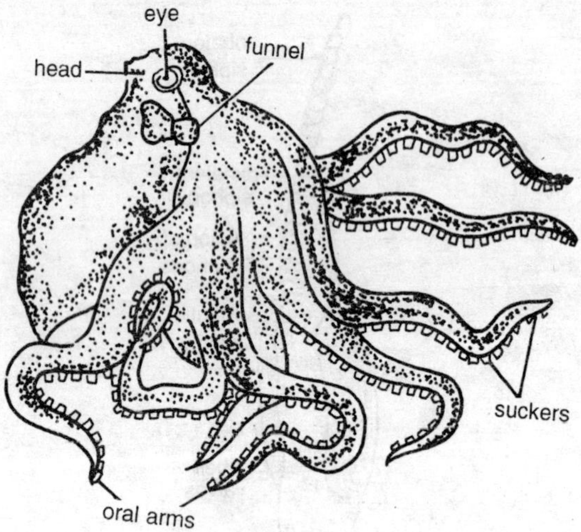

Fig. 1.24. *Octopus.*

8. Third arm on the right side in male modified to act interomittent organ or *hectocotylized*.
9. Like *Sepia* and *Loligo*, *Octopus* is also capable of producing inky liquids which form a smoky screen water.
10. Sexes are separate and there is no metamorphosis as the development is direct. Female cares the eggs.
11. Mouth with sharp and beak-like horny jaws.

ARGONAUTA

Phylum	-	Mollusca
Class	-	Cephalopoda
Subclass	-	Dibranchiata
Order	-	Octopoda
Genus	-	*Argonauta*

1. It is a marine animal found in Atlantic and Indian oceans.
2. It is commonly called as '*paper nautilus*'.
3. Sexual dimorphism is remarkable and well-marked.
4. Male is 2.5 cm in length and without shell.
5. Female is 20 cm length and possesses a thin. transparent shell secreted by the extremities of two dorsal arms.

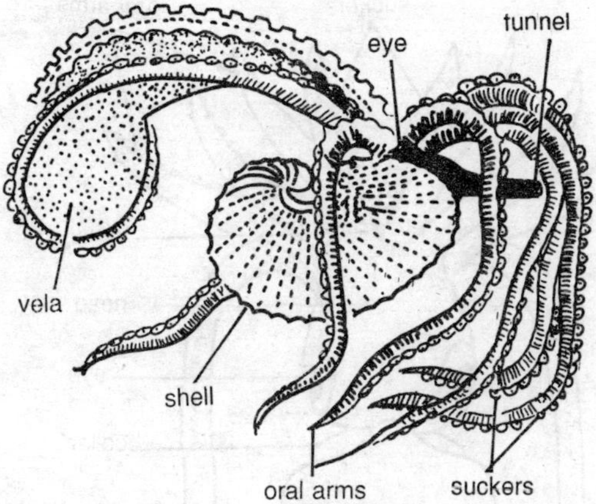

Fig. 1.25. *Argonauta.*

6. Shell is used for protection of eggs.
7. Third arm in male is *hectocotylized*.
8. Eyes and funnel are present.
9. Two plume-like ctenidia and two kidneys are present.

LOLIGO

Phylum	-	Mollusca
Class	-	Cephalopoda
Subclass	-	Dibranchiata
Order	-	Decapoda
Genus	-	*Loligo*

1. *Loligo* is commonly known as *squid*.
2. It is found in warm seas and coastal shallow or deep waters.
3. It has world-wide distribution found along the Pacific and Atlantic coasts, China, India and U.S.A.
4. Body is fleshy, dorsoventrally flattened and differentiated into anterior head and a posterior pointed visceral hump with lateral fins.
5. The colour is dark greyish with raddish spots. It is *metachrotic* having colour changing habit.
6. Head bears a pair of large eyes, and a central mouth surrounded by ten arms.
7. Foot is modified into the funnel and ten arms.
8. Eight arms are short, stumpy and non-retractile, while two are long, slender and retractile tentacles or arms used for capturing the prey.

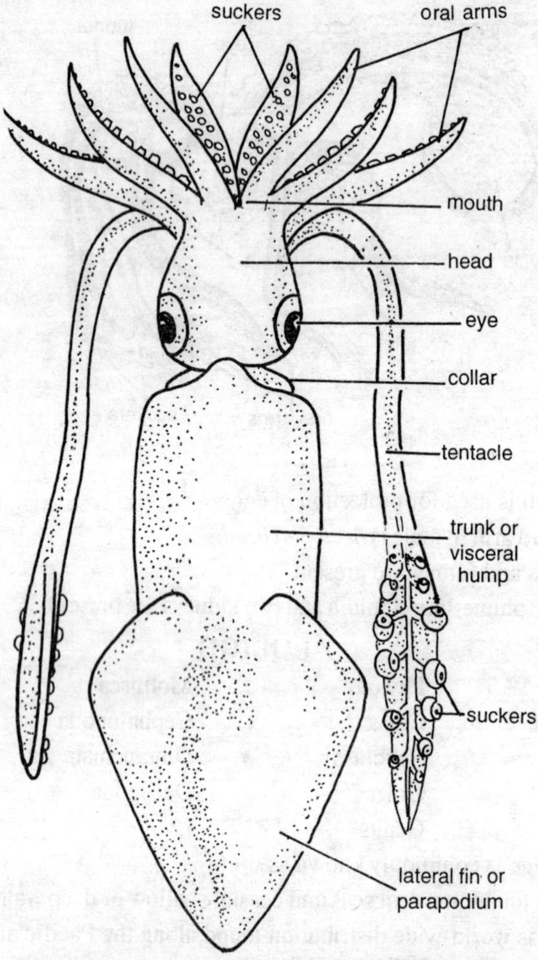

Fig. 1.26. *Laligo*.

9. The inner surface of the tentacles and arms is provided with two rows of suckers.
10. The funnel is muscular tube.
11. Mantle is thick and muscular, encloses the visceral mass and mantle cavity.
12. Shell is internal, feather-shaped plate concealed beneath the mantle on the anterior surface.
13. Two gills and two kidneys are present in the mantle cavity.

14. An ink-sac is present for defence.
15. Sexes are separate. The eggs are deposited in long cylindrical jells masses attached together at one end.
16. It is used as food by Italians and Chinese.

NAUTILUS

Phylum	-	Mollusca
Class	-	Cephalopoda
Subclass	-	Tetrabranchiata
Genus	-	*Nautilus*

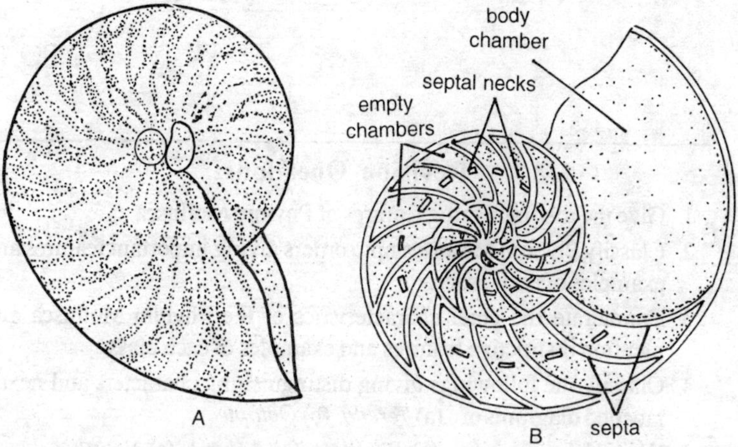

Fig. 1.27. *Nautilus.* A—External Shell, B—Internal Chamber.

1. It a marine cephalopod living in shallow waters and shores and coral reefs in the Indian and South Pacific Oceans.
2. It is commonly known as '*Pearly nautilus*'.
3. The shell is well developed, spirally coiled and composed of calcareous matter, which is divided into several compartments by a series of septa.
4. Through the various compartments runs a tube called *siphuncle* the walls of which are formed of calcium carbonate. A narrow vascular prolongation of visceral region of the animal passes through this tube.
5. Head is well developed bearing the eyes.

6. The anterior part of food is modified into several arms provided with tentacles which are arranged in two series.
7. The posterior portion of the foot is modified into funnel situated below the head.
8. Buccal cavity with a pair of jaws and odontophore bearing several teeth.
9. Two pairs of gills and kidneys are present.
10. They are *hyperpolyandrous* i.e. male are abundant than female.
11. Presence of 60 tentacles in male and 90 in female.
12. Pericardium leads directly to exterior by two orifices located near the posterior renal aperture.

Revision Questions

1. Give the distinguishing features of Phylum Mollusca.
2. Classify Phylum Mollusca upto orders giving important features and examples of each order.
3. Enumerate the chief characteristics of the Phylum Mollusca and mention distinctive features and examples of each class.
4. Classify the followings giving distinguishing characters and neatly labelled diagrams of : (a) *Teredo,* (b) *Octopus.*
5. Write short notes on : (a) *Pincdata,* (b) *Aplysia,* (c) *Nautilus.*

2

Neopilina

Neopilina belongs to class-Monoplacophora of phylum *Mollusca*. Before the discovery of living specimens of *Neopilina,* Monoplacophora had been known only as fossils from lower Cambrian to the Devonian periods. All the monoplacophorans bear a single symmetrical shell, which varies in shape from a flattened shield-like plate to a short cone.

Discovery

In 1952, ten living specimens of *Neopilina* were dredged from a depth of 3500 meters ocean trench off the Pacific Coast of Costa Rica. They were first reported in 1957 by *Lemche* as *Neopilina galatheae.* The detailed structure was described by *Lemche* and *Wingstrand* (1959). Since this initial discovery, specimens belonging to seven different species have been collected from deep water (2000-7000 meters) in various parts of world. *Neopilina ewingi* from Peruchile Trench, *N. veleronis* from Gulf of Aden.

EXTERNAL MORPHOLOGY

Shape and size

The living specimens of *Neopilina* are 3mm. to little more than 3 cms. long and externally resemble a combination of Chiton and gastropods. The body is oval and a *bilaterally symmetrical.*

Shell

There is a single, large, bilaterally symmetrical shell present on the dorsal side. The apex is directed anteriorly. The shell is three-layered made up of an outer *periostracum,* a middle *presmatic layer* and inner *nacreous layer.*

Mantle and mantle cavity

Just within the edge of the shell lies the mantle. It secretes the shell on outer side. A mantle groove or *pallial groove* separates the edge of foot from the mantle on each side.

Foot

A broad, flat, muscular and almost circular foot is present on ventral side. It helps in clinging and creeping. It is a ciliated structure bearing numerous gland cells.

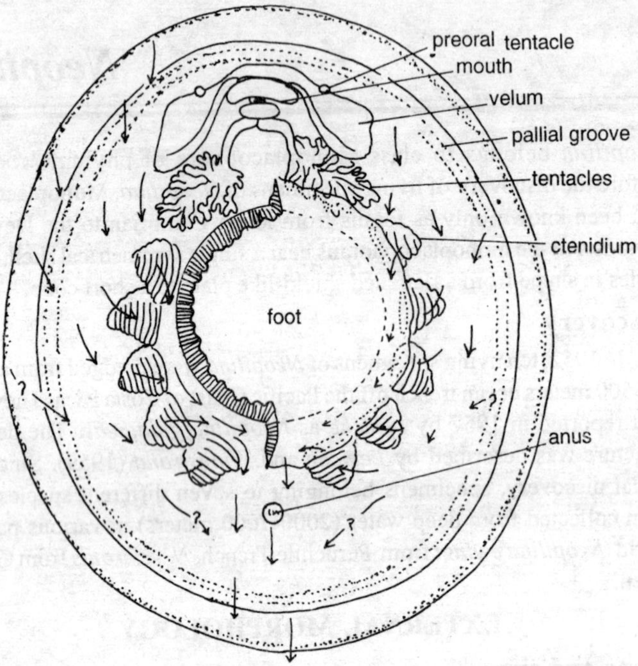

Fig. 2.1. *Neopilina*, a monoplacophoran. Ventral view, with a part of the foot cut away, and the probable course of water current indicated by arrows.

Organs of mantle cavity

The mouth is located in front of the foot, and the anus is located in the pallial groove at the posterior end of body. In front of the mouth is a *preoral fold* or *velum,* which extends laterally on each side as a rather large ciliated palp-like structures. Behind the mouth lies a *postoral fold*, forming a pair of tentacles. The velum and tentacles may be associated with feeding. In *N. galatheae* the pallial groove contains five pairs of unipectinate gills. In *N. ewingi* six pairs of gills are present. Five pairs of minute *nephridiopores* open at the base of the gills while 6th pair opens anteriorly.

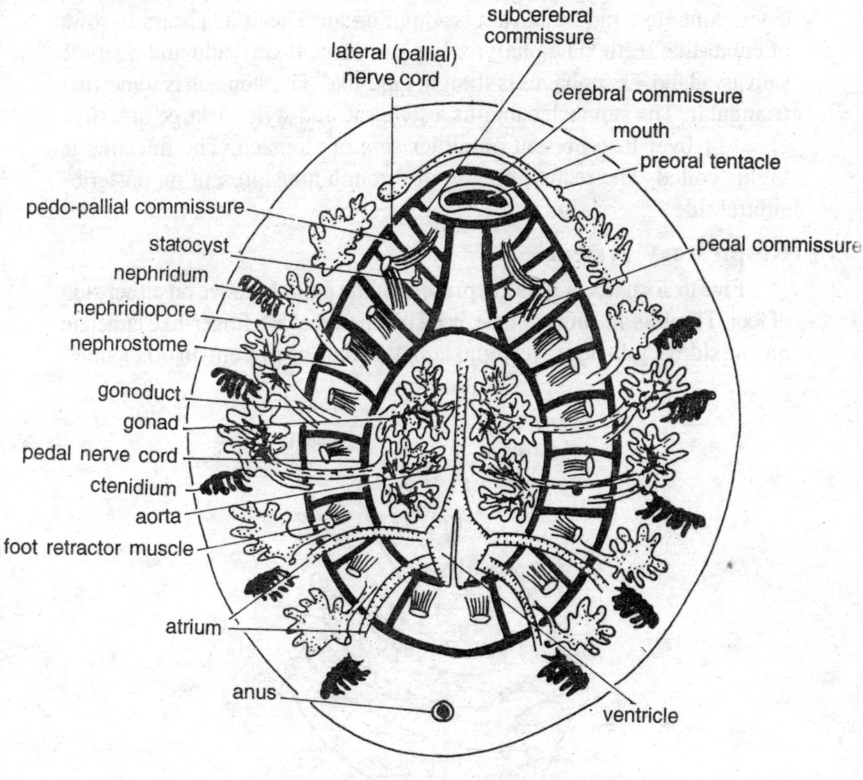

Fig. 2.2. *Neopilina,* a monoplacophoran. Internal organization.

INTERNAL STRUCUTRES

Musculature

Eight pairs of pedal retractor muscles are present. Their dorsal ends are inserted into the interior of the shell. Buccal musculature is complicated.

Coelom

Coelom is made up of three components—*pericardial, dorsal* and *ventral.* The heart is surrounded by a paired pericardial coelom. In front of it, paired dorsal coeloms are present. But both the coeloms have no connection. The ventral coelom is represented by two pairs of sacs forming the gonads.

Digestive system

The alimentary canal is complete. It starts from mouth and terminates into anus. The mouth is present at the antero-ventral side. The buccal cavity contains a radula and a subradular organ. The radula bears 16 rows of comb-like teeth. The pharynx has two large diverticula and a single salivary gland. Oesophagus is straight tube-like. The stomach is somewhat triangular. The stomach contains a style sac and style. A large digestive gland or liver lobe present on either side of stomach. The intestine is highly coiled. The rectum opens out through anus present on posterio-ventral side.

Respiratory system

Five to six pairs of gills are present in the pallial groove on either side of foot. The gills are unipectinate, each composed of 5-8 finger-like lamellae on one side of gill-stem. Vestigial lamellae may be present on other side.

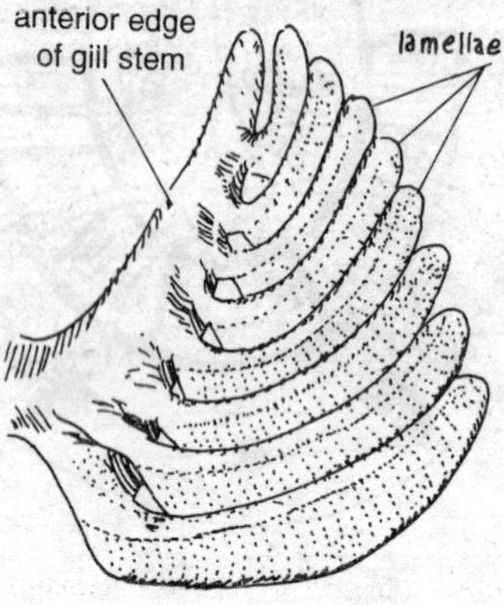

Fig. 2.3. *N. galatheae.* A gill in ventral view.

Circulatory system

The circulatory system includes a heart, vessels and sinuses. The heart is surrounded by a paired pericardial coelom, consists of two pairs of *atria* or *auricles* opening into a pair of *ventricles*. From four *efferent gill vessels* the blood is received into the anterior pair of auricles, the posterior pair receive blood from posterior pair of gills. The two ventricles are divided by the rectum. Each ventricle gives rise to an anterior *aorta*. Both the aortae unite to form the *dorsal aorta* which opens into the head sinuses.

Excretory system

There are six pairs of *nephridia* located on each side of the body, and except for the first pair, each nephridium opens into the coelom (three pairs into the body coelom and two pairs into the pericardial coelom). The nephridiopores open into the pallial groove, the last five pairs located near the gills.

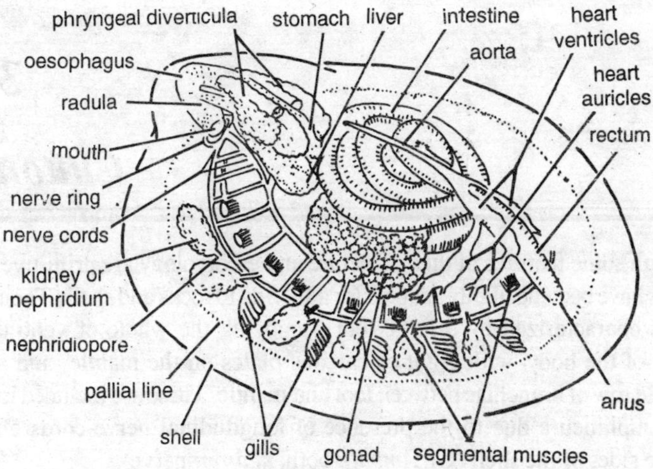

Fig. 2.4. *Neopilina.* Internal anatomy in side view.

Nervous system

There is a circum-oral nerve ring or *cerebral commissure* surrounding the mouth. From the nerve ring arise two pair of posterior longitudinal cords, the outer *visceral cords* and inner *pedal cords.* These unite posteriorly to form loops. Ten pairs of *transverse connectives* connect the pedal and visceral cords. Nerves extend from the cords to the gills.

Reproductive system

The sexes are separate and two pairs of gonads are located in the middle of the body. Each gonad connects through a gonoduct to one of the two pairs of nephridia in the middle region of the body. Fertilization must occur externally. The mode of development is unknown.

3

Chiton

Although some features of their structure and embryology are primitive, Chitons have become highly adapted for adhering to rocks and shells. Their body is characterized by a large foot occupying the whole of ventral surface of the body, with eight calcareous plates on the mantle, and a complete row of branchiae between foot and mantle. These are included in class Amphineura due to the presence of longitudinal nerve cords on both the sides of the shell (Gr. *Amphi*= both, *neuron*=nerve).

SYSTEMATIC POSITION

Phylum	-	Mollusca
Class	-	Amphineura
Order	-	Polyplacophora
Genus	-	*Chiton*

Distribution

Chitons are popularly known as '*sea mice*' or '*Coat-of-mail shell*' are all marine. They are found all over the world at all the depths, though most abundant on the shores between tidal limits. There are approximately 600 existing species of *chiton*. The fossil record, which dates back to the upper Cambrian Period, is rather sparse. Some 350 fossil chitons are known.

Habits and Habitat

The chitons are mostly nocturnal, shy, and sluggish animals. They prefer dark areas during day time but creeping about the rocks at night to feed on diatoms or small encrusting algae, scraped by the use of radula. Thus they are herbivorous. When at rest, they adhere firmly to the under surface of the mantle against the object. When disturbed, they clamp the foot and mantle still more firmly and are then almost impossible to remove without injury. They creep slowly, but when dislodged, they promptly roll up like a pill-bug or millipede, perhaps a defensive mechanism.

EXTERNAL FEATURES

Shape

The body is elongated, elliptical with a convex dorsal surface bearing eight overlapping calcareous plates protective in nature. On ventral side foot, head and visceral mass are present.

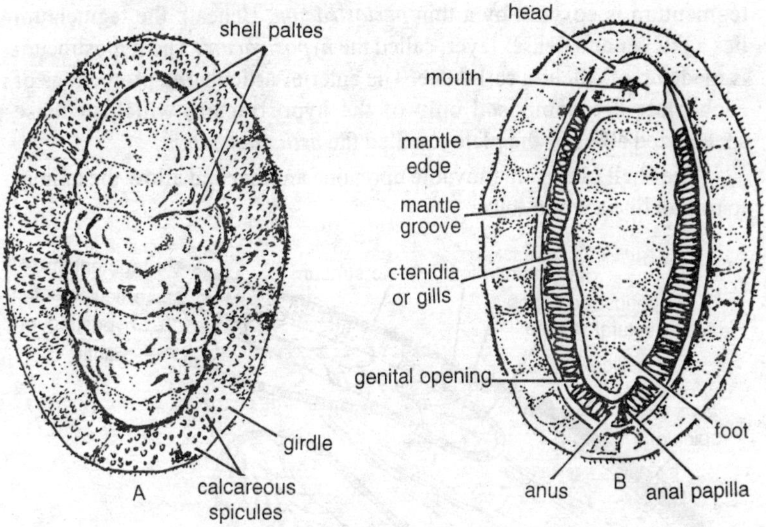

Fig. 3.1. *Chiton*. A—dorsal view, B—Ventral view.

Size

Chiton species range in size from a few centimeters or less. *Chaetopleura apiculata* is one cm. in length found in Atlantic Coast. The giant Pacific species, *Cryptochiton stelleri* measures 30cm. in length. However, most species are 3-12cm. in length.

Colouration

Chitons are usually not brighly coloured; dark shades of red, brown, yellow and green are common.

Shell and Mantle

The most distinctive characteristic of chitons is the shell. The shell is divided into eight overlapping transverse plates. From the nature of the shell is derived the name of the order Polyplacophora. Each plate is similar to the others except the first and last. The first is called *cephalic plate* and the last is *anal plate* which are hemispherical in shape. Except for the overlapping posterior edge, the margins of each plate are covered by mantle tissue. Thus the shell is partially embedded in mantle.

The shell plates are composed of several layers. The upper layer, called the *tegmentum*, consists of an organic conchiolin matrix impergnated with calcium carbonate and is sculptured on its exposed surface. It is perforated by numerous vertical canals containing sense organs. The tegmentum is covered by a thin *periostracum*. Beneath the tegmentum lies a thicker and denser layer, called the *hypostracum*. The hypostracum is made up of calcium carbonate. The anterior articulating projections of each plate are composed only of the hypostracum, which in these specialized areas of the plate is called the *articulamentum*.

The shell plates are movable upon one another and allow the animal to roll up like a wood-louse.

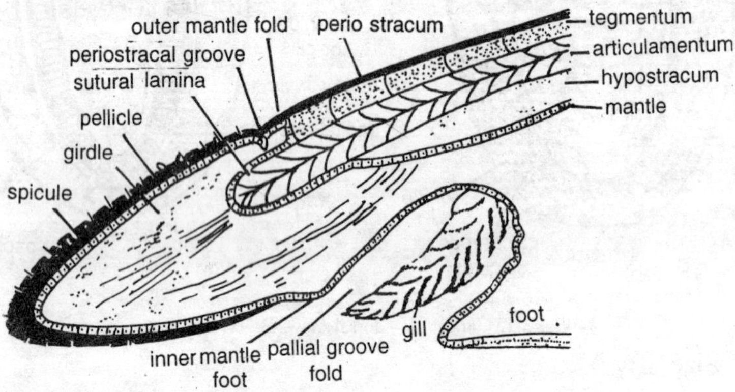

Fig. 3.2. *Chiton*. T.S. through gidle, mantle groove and foot.

The peripheral area of mantle, called the *girdle*, is very heavy and extends a considerable distance beyond the lateral margins of the plates. The girdle surface displays a variety of ornamentation. It may be naked and smooth or covered with scales, bristles or calcareous spicules. The bristles or spicules may be so long and dense that the animal has a shaggy appearance.

Foot and Locomotion

The foot is broad, flat and muscular powerful organ present on the ventral side of the body. It functions in adhesion as well as in locomotion. Adhesion is effected by both the foot and the girdle. The foot is responsible for ordinary adhesion as it is abundantly supplied with a slimy secretion but when it is disturbed, the girdle is also employed.

Chitons are very sluggish animals and may remain in one area for a long period. Most species are motionless at low tide. It moves by employing mucous and waves of muscular activity.

Head

The head is small, cylindrical, inconspicuous structure and is not distinctly marked off from the rest of the body. It contains the month mid-ventrally. The eyes and tentacles are absent. The head is separated from the foot by a narrow groove. The head is provided with a *labial palp* on either side of the month.

Mantle cavity and groove

A space is left in the form of shallow groove between mantle and foot and is called *pallial groove*. The *ctenidia* or gills are lodged in this groove being arranged in a single row on each side. The pallial groove also contains the mid-ventral *anus*, two *renal apertures* one on each side slightly in front of anus and two *genital apertures* one on each side in front of the renal apertures.

Coelom

The body cavity is an extensive cavity and is divisible into two parts, the primary body cavity or haemocoel and the secondary or true body cavity. The haemocoel forms an extensive system of lacunae or sinuses, into which the arteries open. The true coelom is lined by coelomic epithelium and is represented by the pericardium and genital cavities.

DIGESTIVE SYSTEM

The alimentary canal is simple, extends from one end of the body to the other. It comprises:

Mouth

The mouth is situated mid-ventrally on the head leads into a small buccal cavity.

Buccal cavity

It is small, cylindrical and lined with chitin. The lining of buccal cavity is thrown into longitudinal folds. The floor of buccal cavity contains a muscular elevation, the *odontophore*, with a long, horny and ribbon like radula. The radula bears 17 teeth in each transverse row. The lateral teeth are capped with magnetite (iron). The radula is contained in a radula sac above the odontophore. Below the odontophore is a smaller evagination call the subradular sac. Subradular sac contains a cushion-shaped sensory structure, the subradular organ, hanging from the roof. A pair of salivary gland opens through the anterior dorsal wall of the buccal cavity. Periodically the subradular organ is protruded and tastes the substratum again.

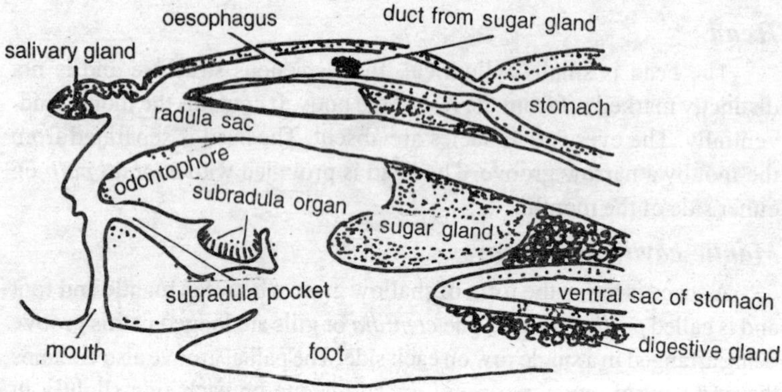

Fig. 3.3. *Chiton.* Buccal region.

Oesophagus

The buccal cavity opens into a short oesophagus which is lined with cilia. It receives ducts of a pair of large, amylase-secreting *pharyngeal glands* or *sugar glands*. The oesophagus leads into stomach.

Stomach

It is an irregularly shaped structure. The wall of stomach is lined generally by long cylindrical cells with coloured granules. It receives the two ducts of a large asymmetrically bilobed and brownish grey *digestive* or *liver glands*, secreting proteolytic and cellulose digesting enzymes.

Intestine

The stomach is followed by long and coiled intestine which may attains a length about four times the body length. It is divided into two sections, an anterior and a posterior parts by a valve. The valve divides the compacted waste, enters the posterior intestine, into short faecal pellets. The valve also slows down the passage of food through the stomach and anterior intestine to allow adequate time for digestion and absorption. The posterior intestine opens out through the anus present in parlial groove behind the foot.

Most chitons are microphagous and feed on fine algae and other organisms that they scrape from the surface of rocks and shells with the radula.

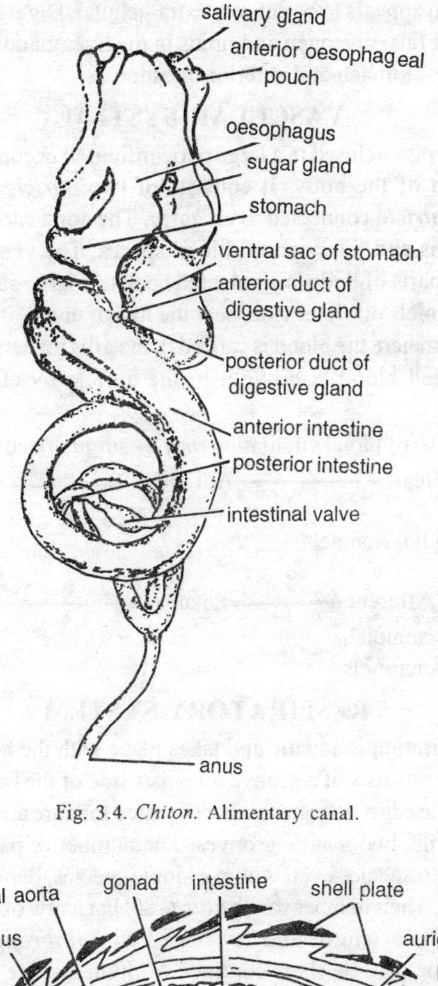

Fig. 3.4. *Chiton.* Alimentary canal.

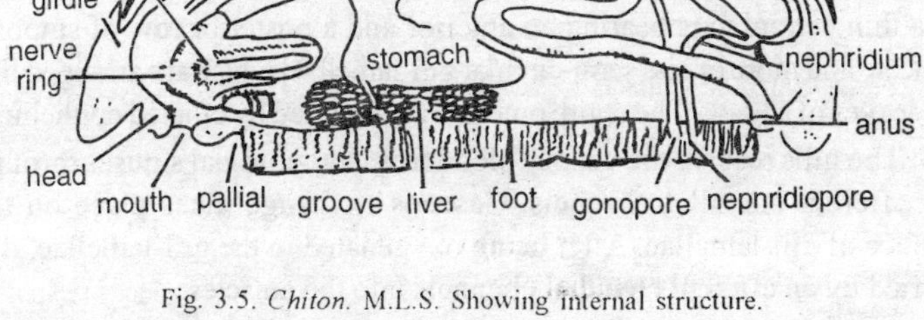

Fig. 3.5. *Chiton.* M.L.S. Showing internal structure.

Digestion appears to be entirely extra-cellular. Digested food is stored in the form of fats, glycogen and lipids in the liver glands. Digested food is absorbed in stomach and anterior intestine.

VASCULAR SYSTEM

The *heart* is enclosed in a large pericardium. It occupies the posterior dorsal region of the body. It consists of two *auricles* and a median elongated *ventricle* connected to an *aorta*. The aorta carries blood to the various organs and intervisceral blood spaces. The venous blood from the different parts of body is conducted back to a large sinus, the afferent ctenidial channels on either side, near the line of union of the mantle with the body. From here the blood is carried to the gills for aeration. From gills the oxygenated blood is brought to the auricle by efferent ctenidial channels.

The course of blood circulation may be summarized as:

→ Heart ⟶ Auricles ⟶ Ventricule
↓
Aorta
↓
Organs of body ⟶ Lacunae ⟶ Afferent ctenidial channels ⟶ Gills
↑
Efferent ctenidial channels → Heart

RESPIRATORY SYSTEM

The respiration is aquatic and takes place with the help of gills. The mantle cavity consists of a groove on either side of the body between the foot and mantle edge. A large number of paired gills are arranged in a linear series within the two mantle grooves. The number of paired gills varies from species to species even within a single species, depending upon the size of animal. Their number varies from 6-80. Each row of gills may extend along the whole length and is termed as *holobranchial type*, eg. *Boreochiton* or may be short confined to limited space at the posterior end of body and known as *mesobranchial type* eg. *Lepidopleurus*.

The gills are inserted at the bottom of pallial groove. Each gill consists of a thin central axis bearing an anterior and a posterior row of smooth, delicate and more or less semi-circular gill-lamellae, which are arranged like the leaves of a book. The entire outer surface is lined by ciliated epithelium.

The gills receive the venous blood from longitudinal sinuses through the afferent ctenidial channels. Gaseous exchange takes place on the surface of gill-lamellae. After being oxygenated in the gill-lamellae, it is carried by an efferent ctenidial channels into the auricles.

On each side of the mantle groove towards the anterior end, the mantle margins are raised to form two inhalent openings. Water enters each inhalent chamber through these anterior openings, runs along the course of groove and passes up through the gills into each dorsal exhalent chambers. The two exhalent currents converge postesiotly and pass to the outside through one or two exhalent openings created by the locally raised mantle.

EXCRETORY SYSTEM

The excretory organs comprise two slender kidneys or nephridia. Each kidney consists of a Y-shaped tube, and provided with numerous

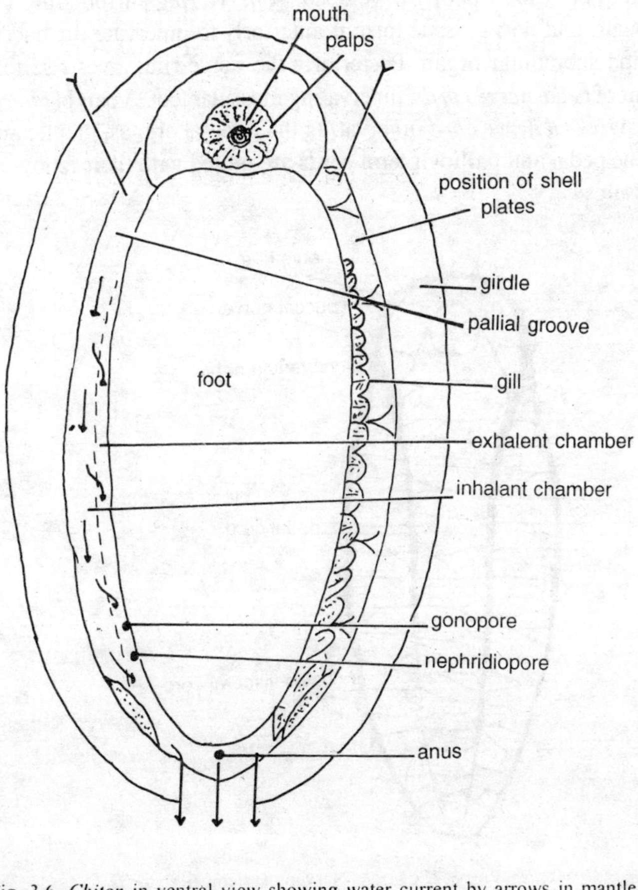

Fig. 3.6. *Chiton* in ventral view showing water current by arrows in mantle groove.

small dendritic outgrowths ramifying along its whole length among the viscera. The tube terminating blindly anteriorly. One of the paired limbs of the Y, the ventral branch, opens to the exterior through a *nephridiopore* into the .posterior part of mantle cavity. The dorsal branch opens into pericardium through a ciliated renopericardial aperture. In this way the pericardial and external apertures of a kidney lie near one another at the posterior end. The excretory surface is increased by numerous branched caeca, that open into all the limbs of the kidneys.

NERVOUS SYSTEM

Chitons possess the simplest and most primitive type of ladder-like nervous system which resembles the turbellarians and trematods. Ganglia are lacking or at least poorly developed. A *nerve ring* surrounding the oesophagus, and nerves issue form it anteriorly to innervate the buccal cavity and subradular organ. Posteriorly the nerve ring gives rise to a large pair of *pedal nerve cords*, innervating muscular foot. A pair of *lateral* or *palliovisceral nerve cord* innervating the visceral orgnas, mantle and shell. The pedal and palliovisceral cords are united with their fellow by commissures.

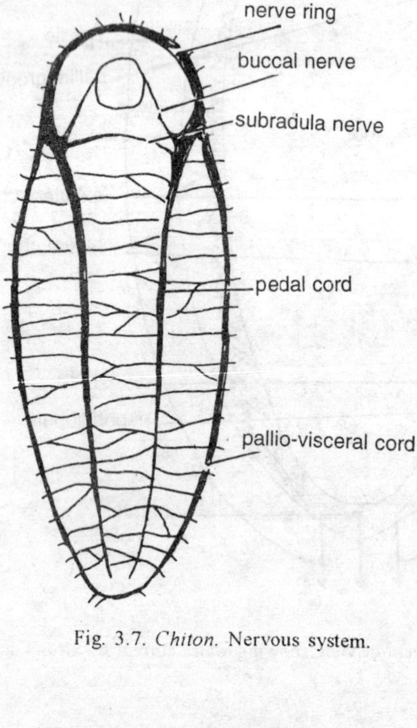

Fig. 3.7. *Chiton*. Nervous system.

SENSE ORGANS

Statocysts, cephalic eyes and tentacles are absent. There are no specialized and well defined sensory organs but according to *Haller*, typical neurosensory cells occur in the ventral epidermis of the head, near the mouth. The chief sensory organs are the subradular organs and the *aesthetes*. Aesthetes, which are unique to chitons, are mantle cells that are lodged within minute vertical canals in the upper tegmentum. The canals and sensory endings terminate beneath a cap on the surface of the shell plates. The function of aesthetes is still uncertain.

In the family Chitonidae, some of the aesthetes are modified as simple eyes are ocelli. The ocelli may be thousands per individual.

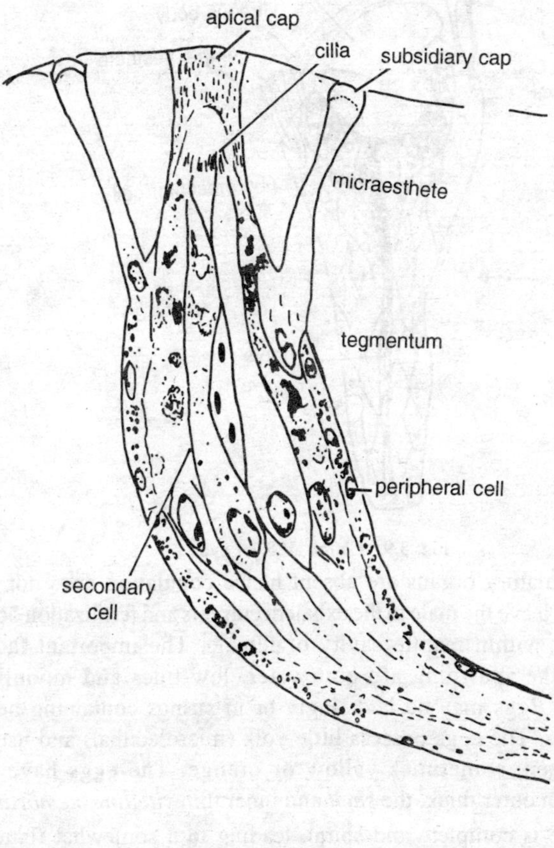

Fig. 3.8. *Aesthete.*

REPRODUCTIVE SYSTEM

Chitons are unisexnal but there is no sexnal dimorphism. The gonads are paired in *Nuttalochiton* but in others the gonad is single and median.

The gonad lies dorsally above the intestine infront of pericardium. The testis and ovary are indistinguishable externally but when the products are mature, the ovary becomes greenish and the testis reddish in colour. Externally the gonad is clothed with a thin layer of muscular and connective tissue. A pair of gonoducts arise from the dorsal surface of the psoterior end of the gonad. They proceed along the anterior wall of pericardium, then curve laterally to open into the mantle cavity infront of the renal apertures.

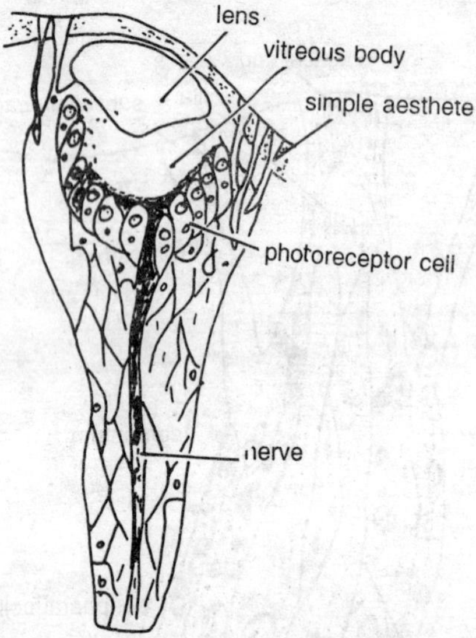

Fig. 3.9. *Chiton.* V.S. of eye.

The copulatory organs are absent hence copulation does not take place. Sperms leave the male in the exhalent currents and fertilization occurs in the sea or within mantle cavity of female. The important factors influencing the spawning are quit water, low-tides and moonlight. Embryology. Eggs may be laid singly or in strings containing nearly 2,00,000 eggs. The eggs possess little yolk (microlecithal) and usually coloured green, sometimes yellow or orange. The eggs have two membranes an outer thick, the *shell* and inner thin *vitelline membrane*.

Cleavage is complete and spiral, leading to a somewhat flattened *coeloblastula*. Gastrulation, mainly of the embolic type formed by the invagination of the macromeres. The blastopore marks the posterior end. The gastrula soon becomes elongated in the direction of the future long axis. The blastopore gradually shifted towards the anterior end of the embryo. The ectoderm around the blastopore is invaginated to form the mouth and stomodaeum. The radular sac arises in the rudimentary from the ventral side of the stomodaeum. By the development of an apical tuft of cilia and a double circlit of cilia or velum, just infront of mouth, the embryo becomes *trochophore larva*. In some cases it becomes free by cracking egg shell whereas in others it remains inside the shell upto the later stage of development.

The trochophore develops an alimentary canal consisting of a large stomodaeum ending blindly since the anus in not yet developed. The anus is formed later by invagination with the proctodaeum. Primitive nephridia, occuring in annelids and many molluscan trochophores, are absent. An apical plate is not present in the early trochophore but it is represented by rudiments of the cerebral ganglia which subsequently appear at the apical pole. The gills appear as papillae and gonads originate by proliferation of the anterior wall of pericardium.

At this stage, the trochophore ruptures the egg envelope and swims for about 15 minutes to several days, depending on species.

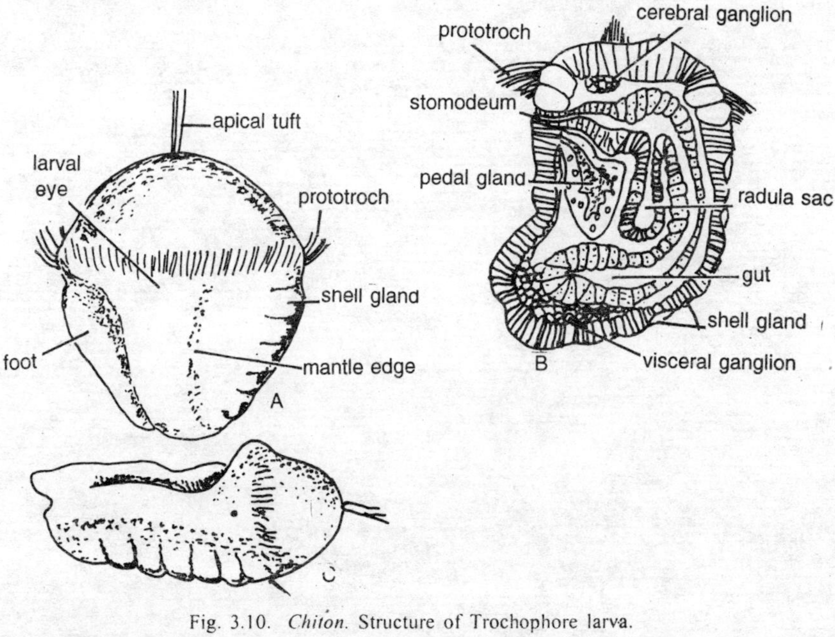

Fig. 3.10. *Chiton*. Structure of Trochophore larva.

Metamorphosis

In the metamorphosis of the *Chiton* trochophore, the post-trochal region elongates to form major part of the body. The shell gland extends up into the pretrochal region, and the shell plates are formed in the transverse grooves. The prototroch degenerates, and the animal sinks to the bottom as a young *Chiton*. The larval eyes are retained for sometimes after metamorphosis.

4

Dentalium

Dentalium is included in class Scaphopoda. Scaphopoda contains about 350 species of burrowing marine molluscs which are popularly known as '*tusk shells*' or '*tooth shells*'. These names are derived from the shape of the shell. The shell is an elongted cylindrical tube usually shaped like a trumpet or elephant's tusk.

SYSTEMATIC POSITION

Phylum	-	Mollusca
Class	-	Scaphopoda
Family	-	Dentalidae
Genus	-	*Dentalium*

Habits and Habitat

These are marine molluscs and world-wide in distribution exept the Polar Regions. They are found all the depths from 6-1830 metres. In India they are commonly found in Mumbai littoral waters and near Gujrat coast. The most common species is *Dentalium octogonum* and rare *D. formosum*. These are found burried in sand or mud. They feed on minute organisms like diatoms, Foraminifera etc.

MORPHOLOGY

Shape and Size

The body is elongted, headless, slightly curved, bilaterally symmetrical, concave dorsally and convex ventrally. The average length is 2 to 5 cm. but *Cadulus mayori* does not exceed 4 mm. in length. *Dentalium vernedei*, found off Japan, is the largest living species, reaching a length of 15 cm.

Colouration

The shells of most scaphopods are white or yellowish, but an East India species of *Dentalium* is a brilliant jet green. *D.formosum* is mottled with shades of dark red.

Mantle

The body is completely enclosed by a tubular mantle, except at an anterior and a posterior apertures. The mantle cavity is large and ventral, extending continuously from one aperture to other. The poserior aperture serves for both inhalent and exhalent water currents.

Shell

The shell is external and secreted by the mantle. It is cylindrical, tubular, slightly curved open at both the ends. The shape of shell is like a trumpet or elephant's tusk. In the genus *Cadulus*, the shell is somewhat cucumber shaped. An operculum is absent and the growth takes place at the larger anterior end. The animal is attached to the shell by muscles near the posterior end.

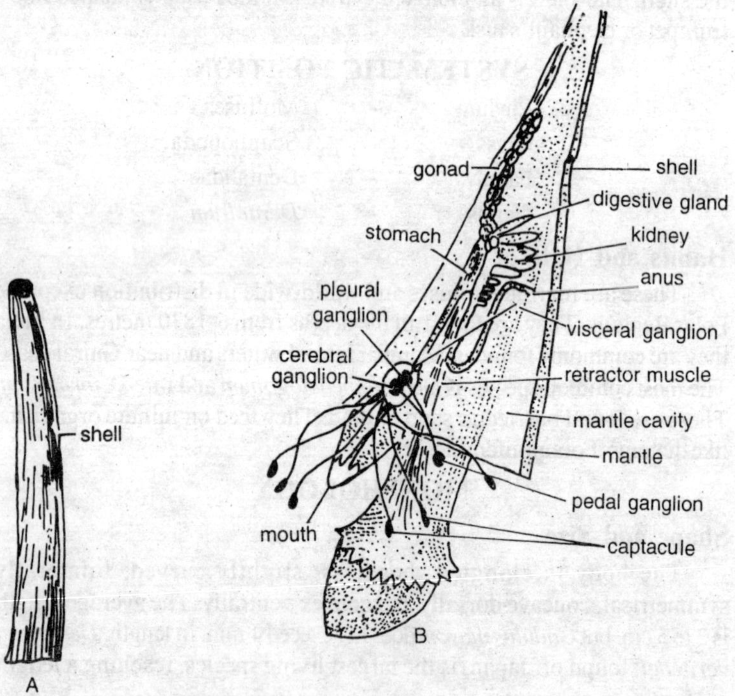

Fig. 4.1. *Dentalium*. A—Shell, B—Structure.

Foot

The foot is long and cylindrical muscular organ, protrudes from the larger anterior end of the shell. It is more or less spade-like and highly extensible. The conical foot is well adapted for burrowing. It is fully extended when gorged with blood, plunges deep into the mud and draws the animal by subsequent contractions. It is pointed in front, but is provided with an oblique wing-shaped fold or pleat on either side of its free extremity. In *Siphonodentalium* the foot terminates in a retractile disc with papillate margins, while in *Pulsellum*, a filiform tentacle arises from the middle of the disc.

Head

Adapted to a burrowing habit, the head is reduced to a short conical projection, or *proboscis*, bearing the mouth. The eyes and tentacles are absent. In *Dentalium*, the mouth is surrounded by a rosette of palp or lobes. On the dorsal side of head, from the two lobes, arise two groups of long ciliated retractile filaments,each having a swollen, sucker-shaped tip. These are called *cephalic filaments* or *captacula*. These are sensory, prehensile, capable of regeneration and serve as tactile organs.

DIGESTIVE SYSTEM

The alimentary canal is somewhat straight and extends from month to anus.

Mouth

It is situated on the short oral proboscis and leads behind into the buccal cavity.

Buccal Cavity

The buccal cavity is dorsally provided with mandible and ventrally a radula and odontophore. The radula is short and provided with five teeth in each transverse row.

Oesophagus

The buccal cavity leads posteriorly into a short oesophagus. It is provided with two large lateral pouches. It leads into stomach.

Stomach

The stomach is simple with a vestigial caecum and gastric shield. It receives the ducts of a large bilobed symmetrical digestive gland or liver.The liver is situated in the upper part of the body.

RESPIRATORY SYSTEM

The respiratory organs such as gills or ctenidia are absent. The respiration is effected by the internal surface of mantle. The mantle is thrown in the form of transverse folds. Cilia are present on ventral folds or ridges. Water enters the mantle cavity by the ciliary action through inhalent siphon and expelled out through exhalent siphon. The water is expelled after 10-12 minutes, due to a violent muscular contraction of body.

BLOOD VASCULAR SYSTEM

The circulatory system is reduced to a system of blood sinuses, and there is a rudimentary heart which is represented by an enlargement of sinus near the anus. The ventricle and pericardium are absent. The blood is colourless and there is no respiratory pigment.

The main sinuses are the perianal, pedal, visceral and pallial.

EXCRETORY SYSTEM

There is a pair of symmetrical nephridia or kidneys, situated infornt of the gonad on the ventral side of the middle region of body, one on either side of the rectum. They open into the mantle cavity by a nephridiopore on either side of the anus. They are without reno-pericardial duct and, therefore, do not communicate with each other.

NERVOUS SYSTEM

The nervous system contains the typical paired cerebral, pleural, pedal and visceral ganglia. The nervous system is symmetrical. The *cerebral ganglia* are situated on the dorsal side of the oesophagus are connected with each other by *cerebral commissure*. An *infra-oesophageal labial commissure* also arises from the cerebral ganglia.

The cerebral and pleural ganglia are connectted to the pedal ganglion of the same side by a long connective which is single at its distal side but bifurcates just before reaching the cerebral and pleural ganglia sending a branch to each. Lying one on either side of the rectum, are ill defined *visceral ganglia*, united by *pleuro-viscaral* connectives to the pleural ganglia. The two pedal ganglia are situateed in foot and are united with each other.

SENSORY ORGANS

There are three differentiated sensory organs:

1. Captacula. These are tactile and olfactory organs, dorsal in position.

2. Subradular organs. It is a ciliated ridge on ventral side of the buccal cavity opposite to the mandible.

3. Statocyst. A pair of statocysts lying in the foot behind the pedal ganglia. Each contains many otoconia.

REPRODUCTIVE SYSTEM

Scaphopods are *dioecious*. The unpaired gonad fills most of the posterior part of the body high up in the visceral dome. The sperms and eggs reach the outside by way of right nephridium. The eggs are shed singly and are planktonic. Fertilization is external.

Development

The cleavage is irregular and takes place just after fertilization. The micromeres multiply much more rapidly than the large macromeres which remains for sometime undivided. The endodermic cells are invaginated to form a gastrula with a large blastopore at its posterior end. The gastrula is succeeded by a free swimming *trochophore larva*. It is a free swimming stage. As a result of further development it is converted into a bilaterally symmetrical *veliger larva*. The head of the larva bears a ciliated girdle or velum consisting of three rings of cilia. Its body shows a rudiment of foot ventrally. A shallow shell gland is formed on the dorsal side and extends ventrally forming two lateral mantle lobes and finally fuse togather forming the tubular mantle round the body.

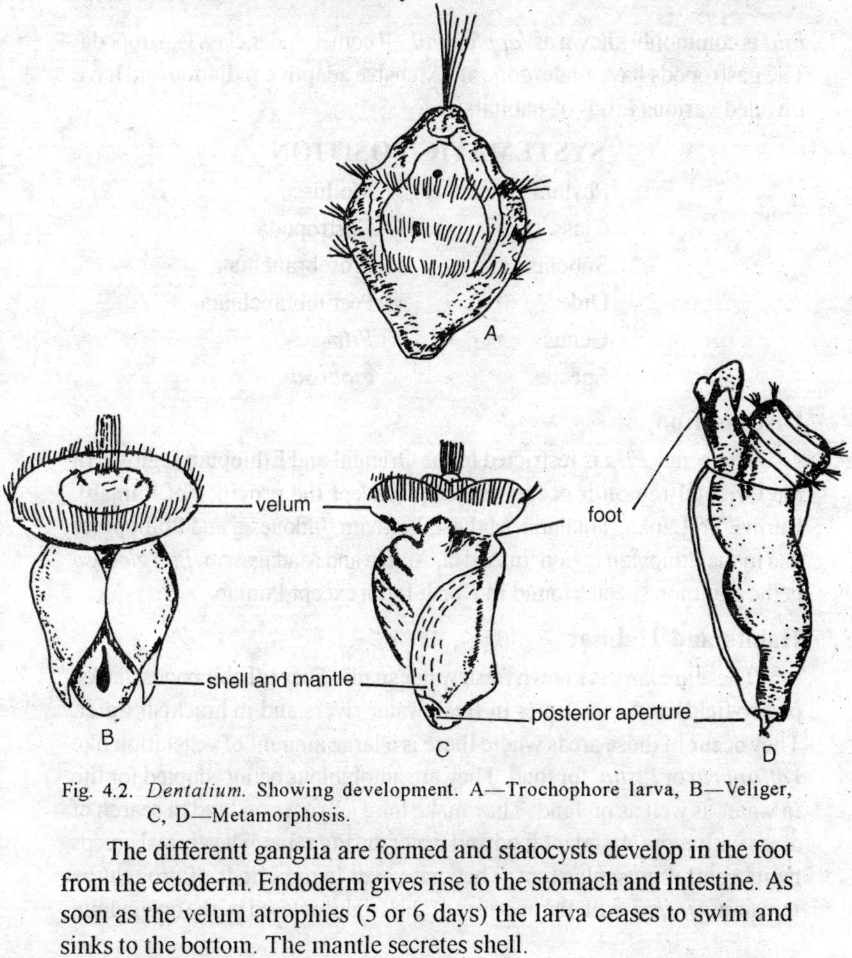

Fig. 4.2. *Dentalium*. Showing development. A—Trochophore larva, B—Veliger, C, D—Metamorphosis.

The differentt ganglia are formed and statocysts develop in the foot from the ectoderm. Endoderm gives rise to the stomach and intestine. As soon as the velum atrophies (5 or 6 days) the larva ceases to swim and sinks to the bottom. The mantle secretes shell.

After sometimes it begins to crawl.

5

Pila globosa

Pila is commonly known as *'apple snail'*. It comes under class Gastropoda. The gastropods have undergone an extensive adaptive radiation and have invaded various kinds of habitats.

SYSTEMATIC POSITION

Phylum	-	Mollusca
Class	-	Gastropoda
Subclass	-	Prosobranchiata
Order	-	Pectinibranchiata
Genus	-	*Pila*
Species	-	*globosa*

Distribution

The genus *Pila* is restricted to the Oriental and Ethiopian regions. In the Oriental region it occurs in India (except the province of Punjab), Burma, Sri Lanka, Thialand, Malaya, Vietnam, Indonesia and Philippines and in the Ethiopian region, in Africa, Arabia and Madagascar. *Pila globosa* is the common species found in North India except Punjab.

Habits and Habitat

These are largest known fresh water snails. They inhabit ponds, tanks, paddy-fields and sometimes in fresh water rivers and in brackish water. They occur in those areas where there is a large amount of vegetation like *Vallisnaria* or *Pistia,* for food. They are amphibious being adapted for life in water as well as on land. They make long journeys on land in search of areas with more abundant food and water during rains. The animal creeps by its ventral muscular foot. They pass over long periods of drought by remaining *torpid,* with the aperture of shell tightly closed by the operculum.

External Features

Shell

The body of *Pila* is enclosed in a calcareous shell, which is somewhat conical and is coiled around a central axis, the coils or whorls being marked off externally from each other by distinct sutures.

The top of the shell is distinguished as the *apex*. Below this are distinguishable the successive lines called *sutures* running a spiral course. The lower most whorl is considerably bigger than all the other whorls taken together, and ends in a wide opening, the *mouth* or *aperture*. In the living animal, the aperture is directed forwards and ventrally and the apex points backwards. The internal cavity contained in the shell, is a continuous one running through all the whorls. The margin of the aperture is named the *peristome*; its outer part is called the *outer lip* and the inner part, the *inner lip* or the *columellar lip*.

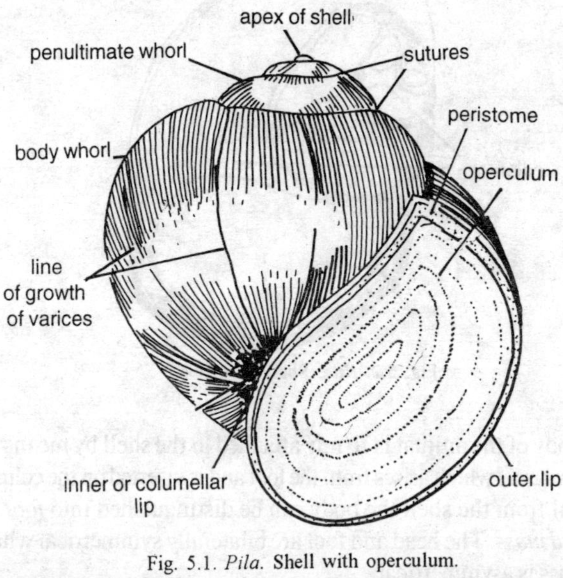

Fig. 5.1. *Pila*. Shell with operculum.

The axis of the shell or the *columella* is hollow and its cavity opens to outside by an opening called *umbilicus*, which lies close to the left corner of the inner margin of the peristome. Into the aperture fits an oblong flat plate, called *operculum* which is composed of material secreted by certain glandular cells of the foot. The inner and outer surfaces of the operculum bear characteristic markings. On its inner surface stands out a sharp elliptic impression called the *boss* which is produced by the attachment of the *opercular muscle*. Round the boss runs a groove. The outer surface of the operculum is marked by a number of lines of growth running round a *nucleus* which is situated subcentrally near the inner margin.

Microscopic Structure of Shell

The shell of *Pila* consists of three layers. The outer layer is known as *periostracum*. It is greenish brown. The underlying two layers are made up of calcareous plates called *ostracum* and *hypostracum*. In the ostracum, the middle layer, the calcareous plates run at right angles to the margin of the aperture of the shell. Ostracum is the thickest of the three layers. In hypostracum, the innermost layer, the plates are arranged parallel to the margin of aperture. The structure of the two inner layers is similar, but the calcareous plates in the two zones lie at right angles to one another.

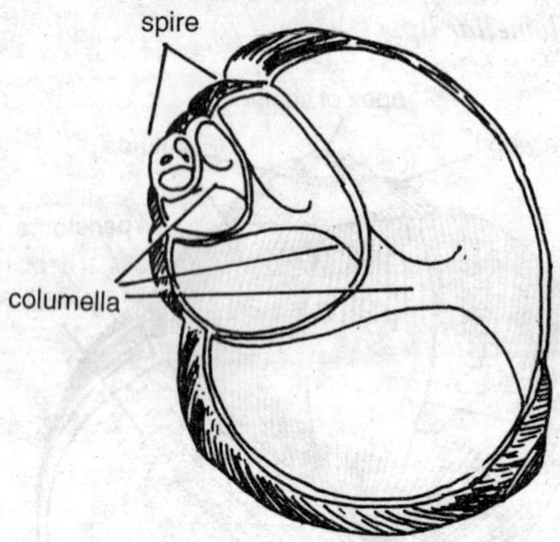

Fig. 5.2. *Pila*. Shell in section.

Body

The body of the animal is firmly attached to the shell by means of the columellar muscle which arises from the foot and is inserted on the columella. On removal from the shell the body can be distinguished into *foot*, *head* and *visceral mass*. The head and foot are bilaterally symmetrical while the visceral mass is asymmetrical.

Head

The anteriormost part of the body, overhanging the foot, forms a definite head. It bears dorsally a pair of small tapering and contractile first pair of *tentacles* or *labial palps*. The snout bears the mouth in its anterior surface below the palps. Behind the labial palps are the longer, filament like true *tentacles* or second pair of *tentacles*. These are hollow and highly contractile and become about 5 cm. long when fully extended, but only 1.5

cm. when contracted. The eyes are placed on tips of short stalks or *ommatophores* which arise from the outer sides of the tentacles, near their bases. On either side of head, connecting it and mantle, there occur two fleshly structures, which are called *nuchal lobes* or *pseudepipodia*. The left nuchal lobe is better developed and forms the respiratory siphon.

Foot

The large, strongly muscular, ventral part of the body forms the foot. When fully expanded, it is roughly triangular in shape, with the apex directed backwards. The foot bears the operculum dorsally on its posterior end, or the *operculiferous lobe*. When the foot is withdrawn, the operculum completly fits into the mouth of the shell and closes it.

The head-foot complex is attached to visceral mass by a short inconspicuous neck.

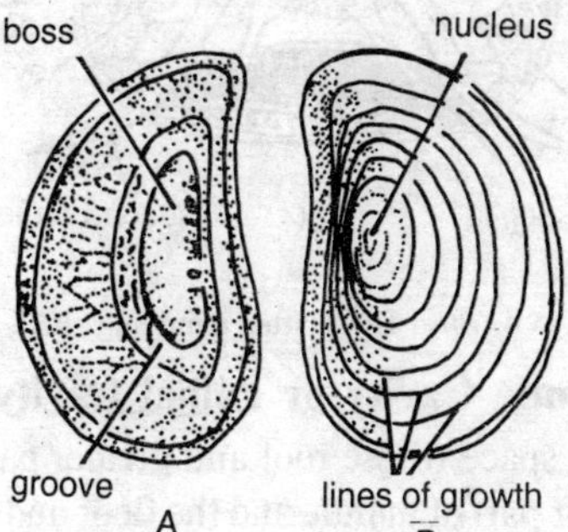

Fig. 5.3. *Pila* Operculum. A—inner view, B—Outer view.

Visceral mass

The visceral mass is a spirally coiled hump-like structure on the dorsal side and contains all the vital organs of the body.

Mantle

The skin of the visceral mass forms the *mantle* or the *pallium*, the characteristic molluscan structure. The free edge of the mantle is thickened and possesses the shell-secreting glands. The mantle keeps a regular check over the anterior part of animal. There is a long, narrow *supramarginal groove* along the thick free edge of the mantle. Behind the supramarginal groove, there is a slight, thickened band, which has the shell-secreting glands.

The shell is secreted by the mantle. The periostracum and the ostracum are secreted by the supra-marginal groove and the shell-glands respectively. The hypostracum is secreted by the general epithelial covering of the mantle, which also serves as an additional organ of respiration.

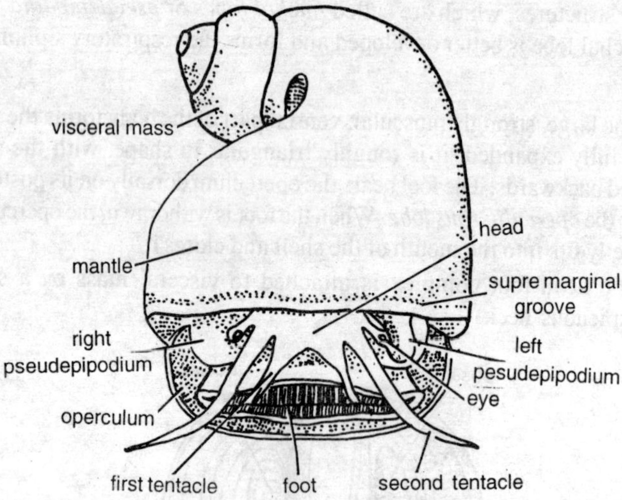

Fig. 5.4. *Pila.* Body after removing shell.

Mantle Cavity or Pallial Cavity

This is the large space whose roof and greater part of the sides are formed by the anterior part of mantle and the floor and remaining part of the sides are formed by the dorsal surface of the body behind the head. In this cavity a number of organs are present known as organs of *pallial complex*. The organs of pallial complex are :

1. Epitaenia

When a longitudinal incision is given along the left of the ctenidium a prominent ridge, the *epitaenia* is seen on the floor of the mantle cavity. It arises near the right nuchal lobe and extends up to the extreme posterior end of the cavity. It divides the mantle cavity into a small right *branchial chamber* and large left *pulmonary chamber.*

2. Ctenidium or gill

It lies at the extreme right side of the mantle cavity hanging vertically downwards from its dorso-lateral wall. Being *monopectinate* (comb-like), its gill filaments or lamellae, which are somewhat triangular in shape, hang freely in the branchial chamber.

3. Rectum

It lies on the left of ctenidium on the floor of mantle cavity and opens to the exterior by anus, situated near the right *pseudepipodium* or right *nuchal lobe*.

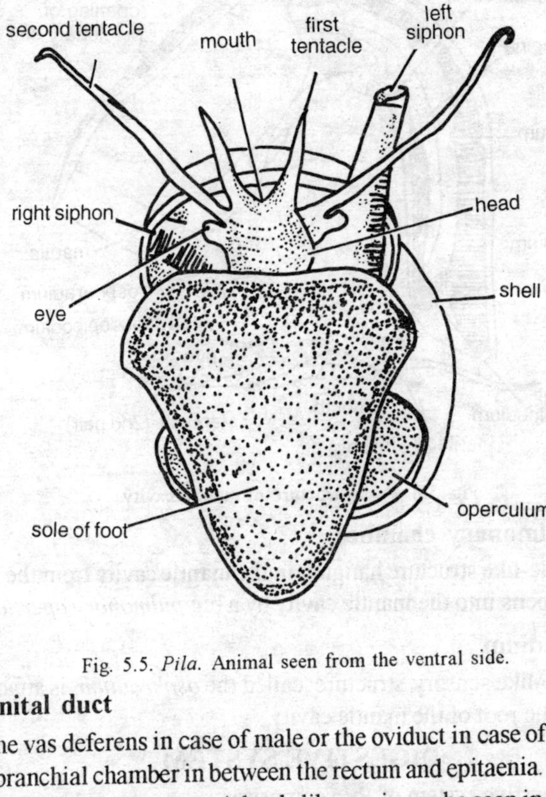

Fig. 5.5. *Pila*. Animal seen from the ventral side.

4. Genital duct

The vas deferens in case of male or the oviduct in case of female lies in the branchial chamber in between the rectum and epitaenia. Its opening is situated close to the anus. A hook-like penis can be seen in front of the genital pore in case of male specimen. At the base of penis is found the hypobranchial gland.

5. Hypobranchial gland

It is a glandular thickening at the base of the penis. Its secretion helps in the act of copulation.

6. Renal opening

The anterior chamber of the renal organ projects into the branchial chamber as a small area of reddish colour, near the posterior termination of the epitaenia. Its external opening is like a transversely oblique slit situated in a shallow depression.

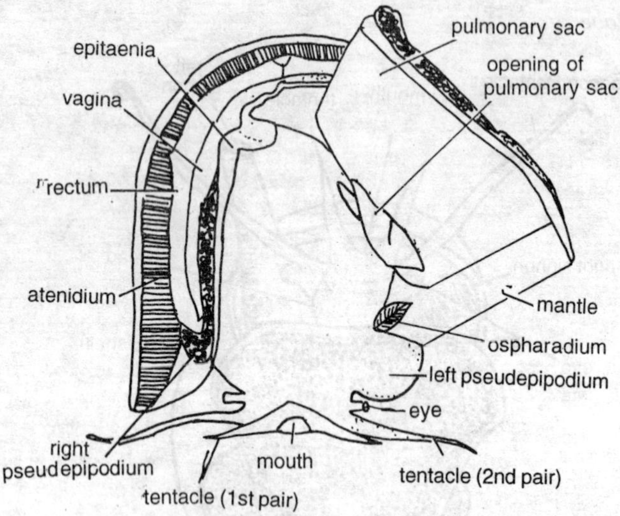

Fig. 5.6. *Pila.* Structure of mantle cavity.

7. The pulmonary chamber

It is a sac-like structure hanging in the mantle cavity from the roof of mantle. It opens into the mantle cavity by a big *pulmonary aperture*.

8. Osphradium

A comb-like sensory, structure, called the *osphradium,* is attached to left side to the roof of the mantle cavity.

DIGESTIVE SYSTEM

The digestive system of *Pila* comprises :

I. Alimentary canal
II. Digestive glands.

I. Alimentary canal

The alimantary canal is tubular and peculiar due to the twisting of visceral mass. It is divisible into three parts, fore-gut, mid-gut and hind-gut.

1. Fore gut

It is also called *stomodaeum* and is ectodermal in origin. It is made up of buccal mass and the oesophagus.

Buccal mass

Mouth is a median vertical slit situated at the tip of snout and is surrounded by frilled folds of integument. True lips are absent. These frills serve as secondary lips. Mouth leads into the buccal cavity, which is the cavity of buccal mass. It is greatly obliterated and contains a pair of jaws and *radula*. The jaws divide the buccal cavity into two sections, the *anterior chamber* also called *vestibule* and a *posterior chamber*. The anterior chamber is extremely short. Its posterior limit is marked by the presence of a pair of bands, which are especially thickened areas of the internal cuticular lining of the buccal cavity. In the posterior region of buccal cavity there lies an *odontophore* or *tongue mass* on the floor of the buccal cavity. On the roof of buccal cavity occur the dorsal *buccal glands*.

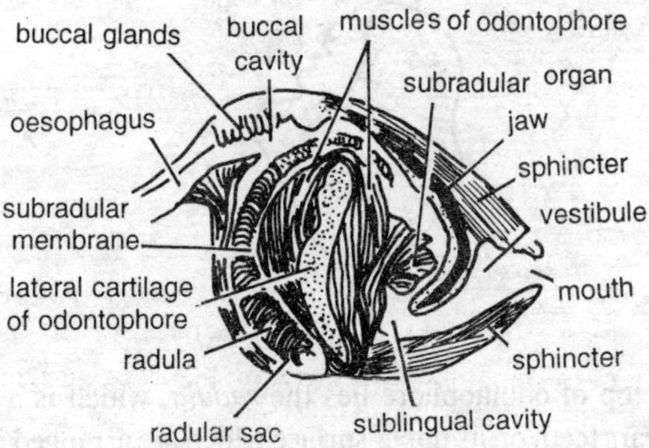

Fig. 5.7. *Pila*. V.L.S. of buccal mass.

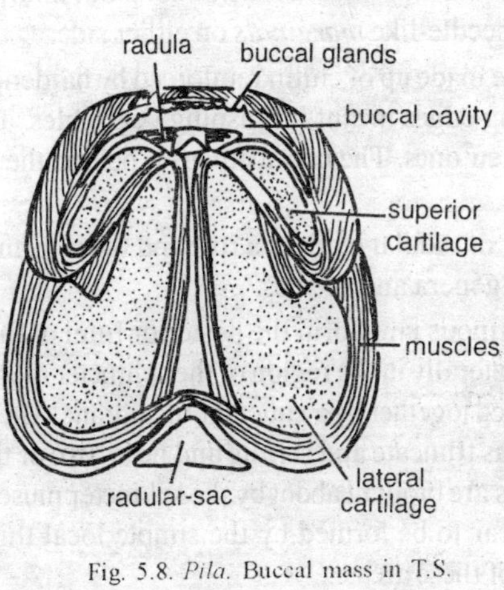

Fig. 5.8. *Pila*. Buccal mass in T.S.

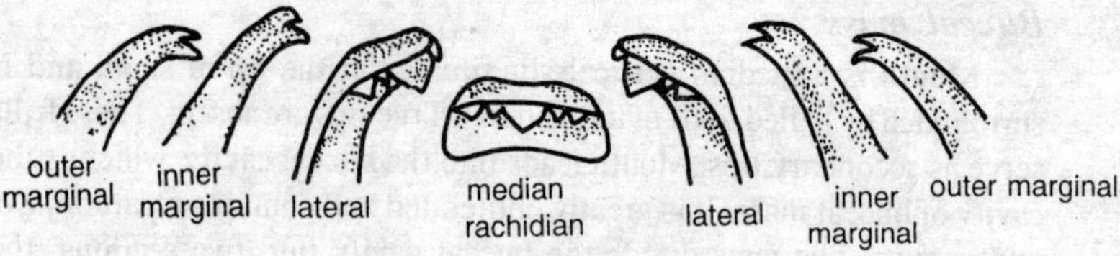

Fig. 5.9. *Pila.* Radula.

The odontophore is supported by *protractor* and *retractor muscles* and two pairs of *cartilages.* A pair of large S-shaped *lateral cartilages,* and a pair of *triangular superior cartilages*. The cartilages project into the buccal cavity.

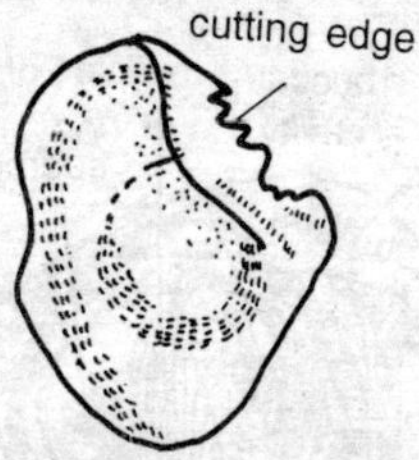

Fig. 5.10. *Pila.* A jaw.

At the top of odontophore lies the *radula*, which is a *star-shaped* body and bears teeth on its upper surface which are arranged in transverse rows as in a file. Each transverse row of radula contains seven teeth including a large *central* or *median rachidian tooth* followed by one small *lateral* and two needle-like *marginals* on either side.

The teeth are made up of chitin reinforced by hardened protein. They bear sharp cutting edges meant for rasping vegetables. The old teeth are replaced by the new ones. The new teeth grow behind the radula and push it forward.

The number of radular teeth in a row and the size and shape of teeth vary in different genera and species.

The two chitinous jaws form the posterior limit of the vestibule. The jaws lying dorsolaterally in the region of the sphincter muscle of the buccal mass are connected together by a cuticular membrane. The anterior cutting edge of each jaw is truncate and serrate and bears two or three large teeth. Their movements are brought about by the sphincter muscles of the buccal mass. They appear to be formed by the simple local thickenings of the cuticular lining of these areas.

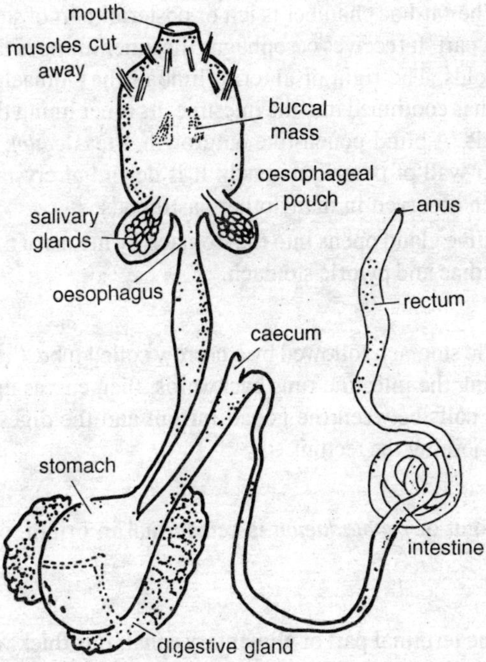

Fig. 5.11. *Pila.* Alimentary canal.

The buccal cavity leads into oesophagus.

Oesophagus

It is like a long and narrow tube. It starts from the dorsal part of buccal mass and extends backwards along the median line to open into stomach after the visceral mass. A pair of round coloured *oesophageal pouches* are found in the lateral sides of oesophagus. Each pouh opens at the junction of the oesophagus and the buccal cavity by means of a narrow duct. The pouches are the projections of oesophageal wall but their function is not well known. Probably they secrete digestive enzymes. They are also used to store food for some times.

2. Midgut

It comprises stomach and intestine. It is endodermal in origin.

Stomach

It lies on the left side of visceral mass, below the pericardium, embedded in the digestive gland. Its internal cavity is U-shaped in appearence. The stomach contains two parts or chambers : the *cardiac* and *pyloric*. The cardiac chamber is left or posterior part of stomach and forms its main part. It receives oesophagus. Its inner lining is thrown into longitudinal folds. The right or anterior limb of the stomach is pyloric chamber which is continued into the intestine. Its inner lining thrown into transverse folds. A blind pouch-like outgrowth, the *caecum*, originates from the lower wall of pyloric stomach. It is devoid of crystalline cone which has been observed in many other gastropods.

The digestive gland opens into the stomach by means of a duct at the junction of cardiac and pyloric stomach.

Intestine

The pyloric stomach followed by a narrow coiled tube, the *intestine*. In the beginning the intestine runs backwards, then curves upward and takes 2½ or 3 coils between the gonad infront and the digestive gland behind, before joining the rectum.

3. Hindgut

The hind-gut or *proctodaeum* is ectodermal in origin, comprising rectum.

Rectum

It forms the terminal part of alimentary canal. It is thick walled tube running into the branchial chamber and opens to exterior through the anus lying about 0.6 mm. from the edge of the right nuchal lobe.

II. Digestive glands

A number of digestive glands are associated with alimentary canal. These are :

1. Salivary glands

Salivary glands are paired, white branching glands lying one on either side of the oesophagus behind the buccal mass. From each salivary gland a duct enters the buccal mass to open into it in the region of the dorsal buccal glands.

They secrete mucous and enzyme. Mucous helps in the lubrication of the radula and the smooth transport of food. The enzyme helps in digestion of starch.

2. Digestive gland

The digestive gland is a large spirally coiled somewhat brownish or dirty-green in colour. It forms the greater part of the visceral mass. It consists or two lobes, the smaller is connected with the stomach, the bigger extends to apex of the spiral of the visceral mass. From each lobe arises a main duct, the two ducts unite to form a single duct before opening into the stomach. These ducts, into the digestive gland, branch repeatedly and end blindly in a very large number of small tubes, the *alveoli*. The alveoli are lined with a digestive epithelium made up of three types of cells - (i) *secretory cells* which secrete a cellulose digesting enzyme, (ii) *resorptive cells* which digest proteins intracellularly, and (iii) *lime cells* or *calciferous cells* which store calcium phosphate. Semidigested food enters into these alveoli, where digestion of cellulose and proteins takes place.

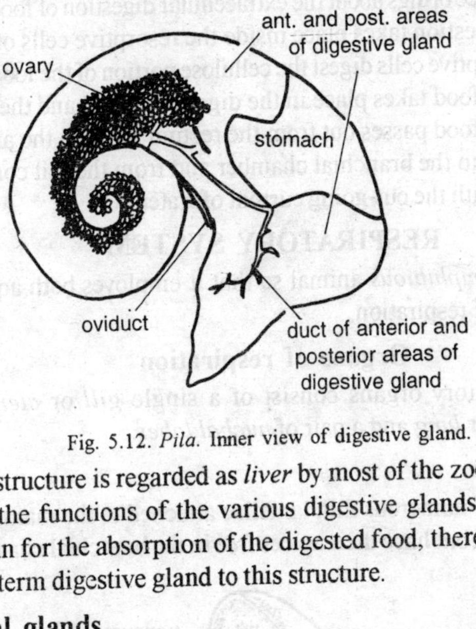

Fig. 5.12. *Pila.* Inner view of digestive gland.

This structure is regarded as *liver* by most of the zoologists but as it performs the functions of the various digestive glands and also is the main organ for the absorption of the digested food, therefore, *Dakin* has given the term digestive gland to this structure.

3. Buccal glands

A pair of elongated glands are present in the roof of buccal cavity. They secrete digestive enzymes.

4. Oesophageal pouches

As already described, they probably secrete digestive enzymes.

Food and feeding

The food consists chiefly of succulent aquatic vegetation. It is also reported that sometimes *Pila* feeds on dead animal tissues too.

By the contraction of the sphincter and protractor muscles of the buccal mass. the jaws move into the mouth and cut thin pieces of plants.

Meanwhile, the radula is brought forward and the pieces of aquatic plants caught by its teeth are carried into the cavity of the buccal mass as the radula moves backwards. The chain-saw movement (described by *Huxley*) of the radula i.e., to and fro movement of the radula over the cartilage just like a band over the pulley is very characteristic. The movement of the radula is very much limited in *Pila* in comparison to other gastropods.

Digestion

The salivary glands pour their secretion into the buccal cavity where it is mixed with the food. It converts insoluble starch into soluble sugar. In the stomach the food is digested by the secretion of the digestive gland. The digestive juice brings about the extracellular digestion of food, whereas intracellular digestion takes place inside the resorptive cells of digestive gland. The resorptive cells digest the cellulose portion of the food. Absorption of digested food takes place in the digestive gland and the intestine. The undigested food passes out from the rectum through the anus. From anus it comes into the branchial chamber and from there it comes to the exterior along with the out-going current of water.

RESPIRATORY SYSTEM

Pila is an *amphibious* animal so that it employes both aquatic and aerial methods of respiration.

Organs of respiration

The respiratory organs consist of a single *gill* or *ctenidium*, a *pulmonary sac* or *lung* and a pair of *nuchal lobes*.

1. Ctenidium

It consists of numerous thin lamellae attached to the mantle wall by their broad bases, but have their apices free in the branchial chamber. The

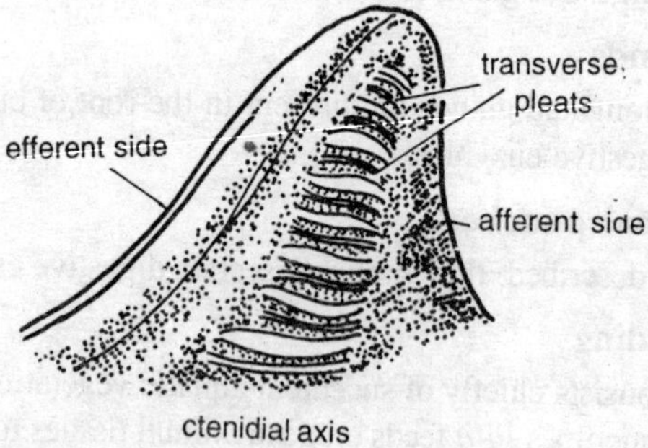

Fig. 5.13. *Pila*. A gill lamella.

line of attachment is known as *ctenidial axis*. A ctenidium with such an axis with a single row of lamellae is described as *monopectinate gill*. The lamellae present in the middle of the ctenidium are largest. They gradually become smaller towards the ends. Each lamella has a longer left side known as *efferent side* being on the side of efferent ctenidia sinus and the relatively smaller right or *afferent side* lying on the side of afferent ctenidial sinus.

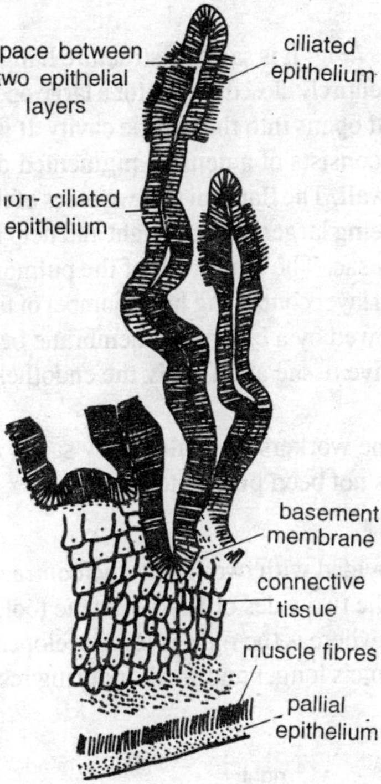

Fig. 5.14. *Pila*. T.S. of two gill lamella.

Shallow transverse ridges, called *pleats* are found in the anterior and posterior regions of each lamellae. Each lamella arises as a double-layer of epithelium internally supported by connective tissue and muscle fibres. The epithelial layer consists of three kinds of cells.

(i) Non-ciliated columnar cells

(ii) Ciliated columnar cells

(iii) Glandular cells.

Underlying the epithelium there is a thin *basement membrane* below which lies connective tissue layer consisting of cells with a few scattered nuclei and large number of obliquely running muscle fibres. After the muscle fibres is situated a layer of pallial epithelium.

The ctenidium of *Pila* is situated on the right side of the body but the distribution of nerves in it and the position of osphradium shows that gills of the left side have been pushed to the right side by extensive development of the pulmonary sac.

2. Pulmonary Sac

It is also known as *lung*. It is sac-like structure hanging from the roof of mantle cavity and is entirely closed except for a large opening (pulmonary aperture) with which it opens into the mantle cavity. It is developed from the mantle itself and consists of a densely pigmented dorsal wall and a creamy white ventral wall. The flaps on the two sides of the opening are of unequal size, the left being larger than the right and help in controlling the opening of pulmonary sac. The outer wall of the pulmonary sac consists of an external epithelial layer containing large number of fine black pigment granules. This is followed by a basement membrane below which lies a thick layer of connective tissue and finally the endothelium of the inner surface.

According to some workers the pulmonary sac is modified second ctenidium but this has not been proved to be correct.

3. Nuchal lobes

The mantle is provided with two fleshy and contractile nuchal lobes or pseudepipodia on the two sides of head over the foot. The left nuchal lobe is well developed whereas the right is less developed. They form two respiratory tubular funnels longer and smaller during respiration.

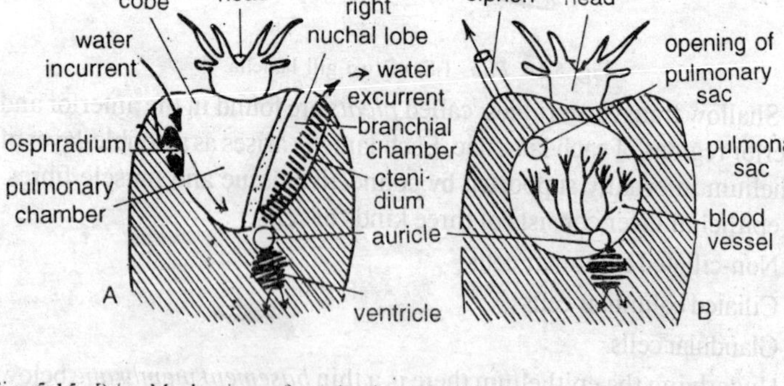

Fig. 5.15. *Pila*. Mechanism of respiration. A—Water current in aquatic respiration. B—Air current in aerial respiration.

Mechanism of respiration

Aquatic respiration

Aquatic respiration usually takes place, when the animal lies submerged in the water, attached to the plants or floating in muddy water. During this process, the animal is fully expanded, and the two nuchal lobes, which are equally expanded, appear more or less like shallow channels. The left nuchal lobe after some time increases in size and takes the form of a gutter. The respiratory current of water enters through the gutter-like left nuchal lobe flowing beneath the *osphradium* and reaches the posterior part of the pulmonary chamber of the mantle cavity and then crossing the *epitaenia* it enters the branchial chamber, where it bathes the entire length of the gill allowing exchange of gases and, finally, passes outside through the *right nuchal lobe.*

Aerial or pulmonary respiration

The aerial respiration occurs when the pond water is poor in oxygen or *Pila* is making excursion on land.

Pila moves towards the surface of water and thrusts out its left nuchal lobe which becomes elongated and tubular to form the *respiratory siphon*. The pulmonary chamber becomes greatly enlarged and is completely cut off from the branchial chamber. The air enters the pulmonary sac through respiratory siphon and the respiratory movements produced by the alternate contraction and expansion of the pulmonary sac help in breathing.

While *Pila* is on land, the respiratory siphon is not formed and the air enters pulmonary chamber directly through the expanded nuchal lobe.

BLOOD VASCULAR SYSTEM

The blood vascular system is highly developed due to its amphibious nature. It consists of :

(I) Heart

(II) Blood vessels

(III) Blood

I. Heart

It is two chambered consisting of an auricle and a ventricle and lies enclosed in pericardium.

(i) Pericardium

The pericardium is thin walled, oval chamber enclosing the heart and situated between the pulmonary chamber and the posterior renal sac on the left side of the body whorl. Posteriorly, it is surrounded by the stomach and the digestive gland. The renal sacs are present on its dorsal side while the oesophagus is found on its ventral side. The pericardial cavity is the true coelom as it communicates with renal organs and the reproductive organ develops from its wall in early stages of development. Pericardium communicates with the posterior renal chamber by *reno-pericardial aperture*.

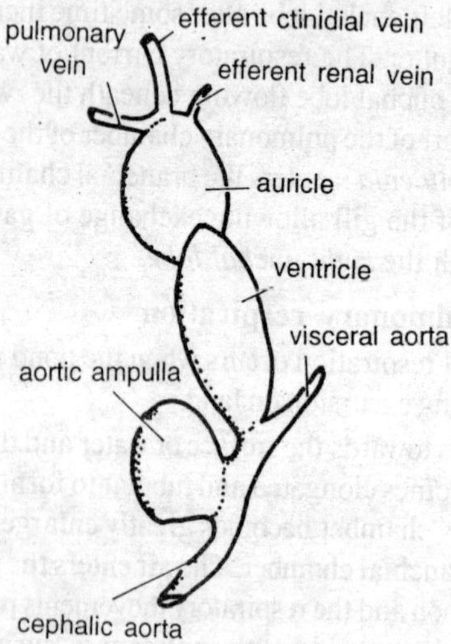

Fig. 5.16. *Pila*. Heart.

(ii) Auricle

The auricle is a thin walled, roughly triangular sac highly contractile in nature. It is situated on the dorsal side of pericardium. It receives blood through three main veins - (a) efferent ctenidial vein, (b) efferent renal vein and, (c) pulmonary vein. At the lower end it is connected with ventricle through the *auriculo-ventricular aperture*. The aperture is guarded by two *semilunar valves* which allow the blood to flow from auricle to ventricle but not in reverse direction.

(iii) Ventricle

It is an ovoid sac lying below the auricle. It has thick, spongy and muscular walls and a reduced cavity due to a coarse meshwork of muscular strands. From the lower end of the ventricle arises a large artery, the *aortic trunk*. The opening between the ventricle and aorta is guarded by two *semilunar valves* which prevent the flow of blood back into the ventricle.

II. Blood Vessels

It includes arteries, sinuses and veins.

1. Arteries

The aortic trunk divides immediately into two branches : an anterior cephalic aorta, and a posterior visceral aorta.

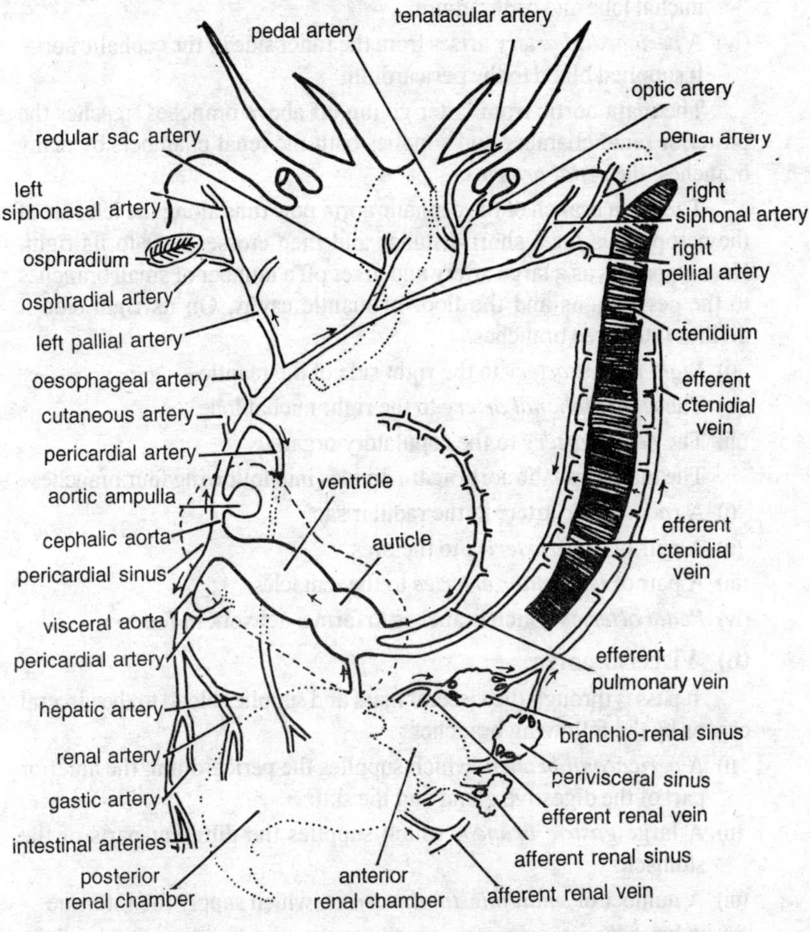

Fig. 5.17. *Pila*. Heart and blood vessels.

(a) Cephalic aorta

There is a thick-walled ampulla at the base of the cephalic aorta. It is highly elastic and helps in pumping the blood. The cephalic aorta, beyond the ampulla, gives off immediately three arteries on the outer side and one artery on inner side.

(i) A *cutaneous artery* which supplies the skin.

(ii) A pair of *oesophageal arteries* which supply the oesophagus.

(iii) A thick *pallial artery* which supplies the left side of mantle, left nuchal lobe and osphradium.

(iv) A *pericardial artery* arises from the inner side of the cephalic aorta. It supplied blood to the pericardium.

The main aortic trunk after giving off above branches, reaches the posterior renal chamber and supplies both the renal chambers by many branches, the *renal arteries.*

The main branch of the cephalic aorta now runs along the left side of the oesophagus for a short distance and then crosses over to its right. Here it appears as a large artery and gives off a number of small branches to the oesophagus and the floor of mantle cavity. On its right side it divides into three branches.

(i) *Right pallial artery* to the right side of the mantle.

(ii) The *right siphonal artery* to the right nuchal lobe.

(iii) The *penial artery* to the copulatory organ.

The main cephalic aorta again divides into following four branches :

(i) A *radular sac* artery to the radular sac.

(ii) A pair of *optic arteries* to the eyes.

(iii) A pair of *tentacular arteries* to the tentacles.

(iv) *Pedal arteries* which branches to form a network in foot.

(b) Visceral aorta

It passes through the visceral mass and supplies blood to the visceral organs by the following branches :

(i) A *pericardial branch*, which supplies the pericardium, the anterior part of the digestive gland and the skin.

(ii) A large *gastric branch*, which supplies the different parts of the stomach.

(iii) A number of *small intestinal arteries*, which supply the intestine.

(iv) A large *digestive branch*, which supplies the digestive gland and the gonad.

(v) Many *renal arteries* to the anterior and posterior renal chambers.

At the end the visceral aorta terminates into branches which supply the rectum and anus.

2. Sinuses

The ultimate branches of the arteries open into wide spaces without walls, called the *blood sinuses*. Some of them act as blood capillaries and contact the arteries with the veins. Other sinuses join together to form larger sinuses. In *Pila* there are four sinuses :

(i) *Perivisceral sinus.* It occupies the anterior part of the body above the foot and below the floor of mantle cavity. It surrounds the anterior part of alimentary canal.

(ii) *Peri-intestinal sinus.* The peri-intestinal sinus is confined around the coils of the intestine.

(iii) *Branchio-renal sinus.* It lies along the right side of the anterior renal chamber.

(iv) *Pulmonary sinus.* It is confined to the walls of the pulmonary sac.

3. Veins

The veins carry the venous blood from the various part of the body to the auricle directly or through the respiratory and excretory organs.

(i) *Afferent ctenidial vein.* It collects blood from the perivisceral sinus, branchio-renal sinus, rectum and the terminal part of the genital duct and carries it to the ctenidium, where it branches to supply blood to the *branchial lamellae* for purification.

(ii) *Efferent ctenidial vein.* It collects oxygenated blood from the lamellae of ctenidium, mantle and copulatory organ and carries to the auricle.

(iii) *Afferent renal vein.* It receives blood from the peri-intestinal sinus and empties into the posterior renal chamber.

(iv) *Efferent renal vein.* It originates from the posterior renal chamber and empties into the auricle.

(v) *The pulmonary vein.* It collects blood from the pulmonary sac and carries it to the auricle.

III. Blood

The blood is colourless and contains colourless *amoebocytes* in the plasma. The blood of *Pila* has a light bluish tinge. A respiratory pigment, the *haemocyanin* is found dissolved in the plasma. The haemocyanin helps in respiration.

Course of Circulation

The ventricle pumps the mixed blood through branches of cephalic aorta and visceral aorta to various parts of the body. The venous blood is collected from the various parts into the perivisceral and periintestinal sinuses. From these sinuses the blood is sent back to the auricle either directly through kidneys for the removal of nitrogenous waste products or indirectly to the gill and mantle for aquatic respiration. The auricle thus gets aerated and non-aerated blood from these organs and it is the mixed blood which is forced into the ventricle to be pumped again to various organs of the body.

EXCRETORY SYSTEM

In *Pila globosa,* there is a single large *renal organ* or *kidney* or *organ of Bojanus* lying behind the pericardium. The excretion is done by the left-kidney as the right kidney has modified into the gonoduct. On one side it opens into the pericardial cavity and on the other, to the outside through the mantle cavity.

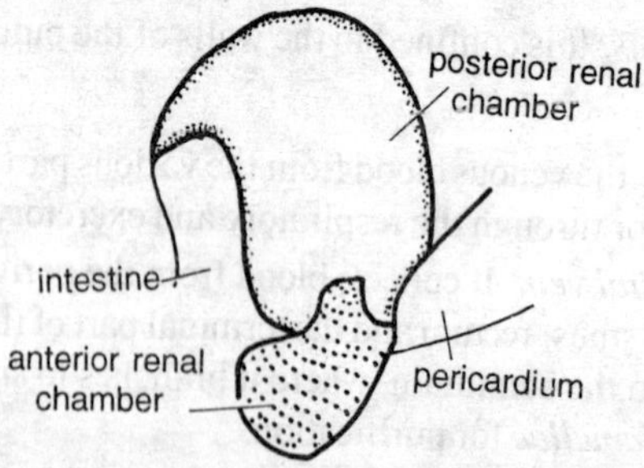

Fig. 5.18. *Pila.* Renal Chamber.

The renal organ is made up of two distinct parts : a right anterior renal chamber and a left posterior renal chamber.

1. Anterior renal chamber

It is an oval reddish organ, lying on the dorsal side of the animal infront of the *pericardium* and *posterior renal chamber.* The cavity of the anterior renal chamber is very small due to the presence of a number of triangular leaf-like processes or *lamellae.* The lamellae are arranged on either side along a thick median longitudinal axis, or the *efferent renal sinus.* On the floor of the chamber, the lamellae are arranged on either side of a median axis, the *afferent renal sinus.* The lamellae arising from roof alternating with the lamellae of arising from floor. The dorsal surface of chamber is marked by numberous transverse grooves, corresponding to these internal lamellae. The anterior renal chamber communicates with the posterior renal chamber through a small opening at one end, and opens by an elongated opening into the mantle cavity lying to the right of the epitaenia, at the other end.

2. Posterior renal chamber

It is a hook-like 'broad chamber' brown or grey in colour situated behind the anterior renal chamber, in between the rectum on the right and the pericardium and the digestive gland on the left. Its large internal cavity encloses a part of the genital duct and a few coils of the intestine. At one end, it communicates with the anterior renal chamber through an aperture, and at the other with the pericardium through an elongated slit-like *reno-pericardial aperture*, perforating a thin vertical *reno-pericradial septum* separating the two. The afferent and efferent renal vessels profusely branch in the roof of this chamber.

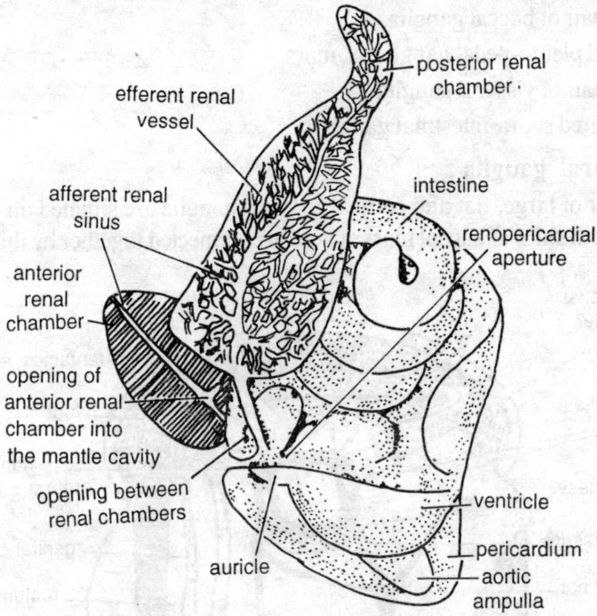

Fig. 5.19. *Pila.* Excretory System.

Physiology of excretion

The renal organ separates the nitrogenous waste products from the blood. The excretory fluid passes from the posterior renal chamber to the anterior renal chamber from where it is discharged into the mantle cavity through the external renal aperture and finally it goes out with the outgoing current of water through right siphon. The excretory matter contains chiefly ammonia, ammonium compounds like urea and uric acid. For conserving water, *Pila* converts ammonia into insoluble uric acid.

In most of the gastropods, the digestive gland also functions as organ of excretion. The wastes are eliminated by way of stomach and intestine.

NERVOUS SYSTEM

A well-developed nervous system is present in *Pila*. It possesses two characteristics (i) except the visceral ganglia all other ganglia form a ring-like structure around the buccal mass (ii) due to torsion the visceral loop is twisted to form the shape of '8'. The nervous system of *Pila* includes the following ganglia, their commissures and connectives:

1. One pair of cerebral ganglia.
2. One pair of buccal ganglia.
3. Paired pleuro-pedal ganglionic mass.
4. One pair of visceral ganglia.
5. Unpaired supra-intestinal ganglion.

1. Cerebral ganglia

A pair of large, flat and conical cerebral ganglia are situated on the dorso-lateral sides of buccal mass. These are connected together by thick,

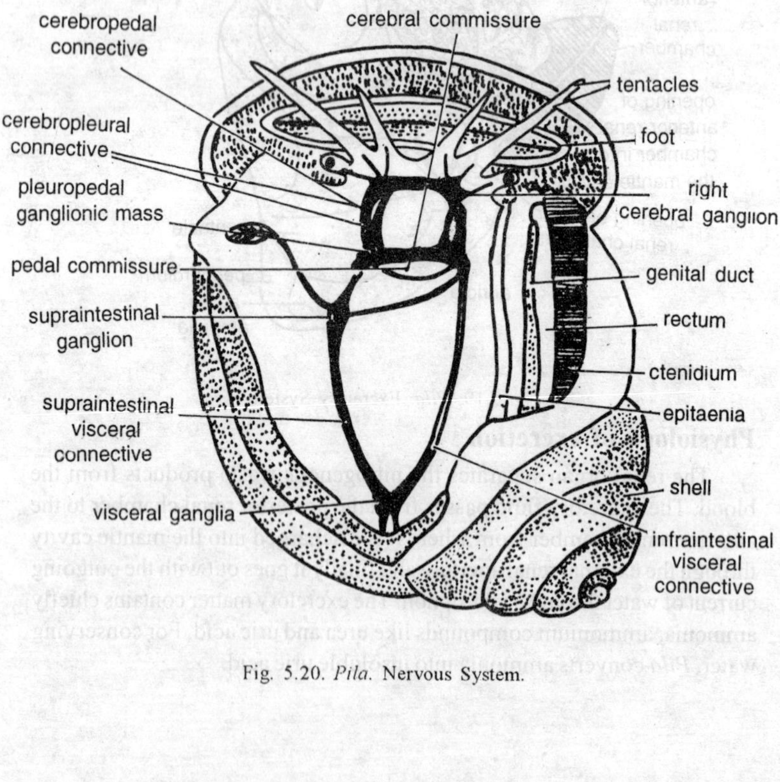

Fig. 5.20. *Pila*. Nervous System.

band-like *cerebral commissure* dorsally. In addition to this both the ganglia are also connected with each other by thin *labial commissure* lying below the cerebral commissure.

Each cerebral ganglion is connected with the buccal ganglion of its side by *cerebro-buccal connective* and by pedal and pleural ganglia by a thick band-like *cerebropedal* and *cerebro-pleural connectives*. All these connectives are situated on the lateral sides of buccal mass.

Anteriorly the cerebral ganglia give nerves to the skin of snout, tentacles and buccal mass. Posteriorly, they give nerves to tentacles, eyes and the statocyst.

2. Buccal ganglia

They are two in number. They are situated on the dorso-lateral sides at the junction of the buccal mass and oesophagus. The two buccal ganglia are connected with each other by a stout *buccal commissure*, which runs on the ventral side of the oesophagus behind the buccal mass. Each ganglion is also connected with the cerebral ganglion of its side by the *cerebro-buccal connective.*

The nerves are supplied from each buccal ganglion to the posterior part of buccal mass and radular sac.

3. Pleuro-pedal ganglia

A pair of large, somewhat triangular ganglionic masses present one on either ventro-lateral side of the buccal mass. Each one is formed by the fusion of an outer *pleural* and an inner *pedal ganglion*, separated by a faint notch. The right pleuropedal ganglionic mass also consists of the *infra-intestinal ganglion* fused with it.

The two pedal ganglia are connected with each other by two *pedal commissures* which are placed one above the other. These commissures are situated below the buccal mass. Many nerves are given off to the foot from the anterior and posterior parts of the pedal ganglion. Behind the pedal commissures, the two pleural ganglia are connected with each other by *infra intestinal nerve*. The left pleural ganglion gives off nerves to the mantle, osphradium, left nuchal lobe, columellar muscles and the anterior part of gill. The right pleural ganglion supplies nerves to parietal wall, epitaenia, right nuchal lobe, copulatory organ, rectum, genital duct and columellar muscles.

4. Supra-intestinal ganglion

It is a fusiform ganglion situated on the left side 1/4th of an inch way from the pleuro-pedal mass of its side. It is connected anteriorly to the left pleural ganglion by *zygoneury* and posteriorly to the left visceral ganglion by thick nerve called *supra-intestinal visceral connective* and on its inner side to the sub-intestinal ganglion of right side by a thin nerve. The thin nerve is known as *supra-intestinal nerve*. On its outer side the supra-intestinal nerve issues a short but thick nerve to the mantle and the anterior part of the ctenidium.

5. The visceral ganglia

It is formed by the fusion of two spindle-shaped ganglion. It lies at the base of the visceral mass. It is connected with the right pleural ganglion by a thin nerve, the *infra-intestinal visceral connective* and with supra-intestinal ganglion by the *left supra-intestinal visceral connective*. It supplies nerves to the renal organ, genital organ and to the stomach and digestive gland.

SENSE ORGANS

The entire exposed surface of *Pila* is provided with sensory cells but followings are special sensory organs :

1. Osphradium

A small osphradium is present in aquatic gastropods while absent in terrestrial snails. It was specially described by *Spengel* (1881), *Bernard* (1890) and *Demal* (1954). It is unpaired in *Pila* and situated dorsoventrally on the roof of mantle cavity. It is bipectinate structure hanging like a curtain in the course of the respiratory water current. It consists of a specialized and usually elevated and ciliated region of the epithelium in which there is an accumulation of the sensory cells. The leaves of osphradium are arranged on both the sides of a central axis. Each leaf is just like a double fold of the wall of central axis.

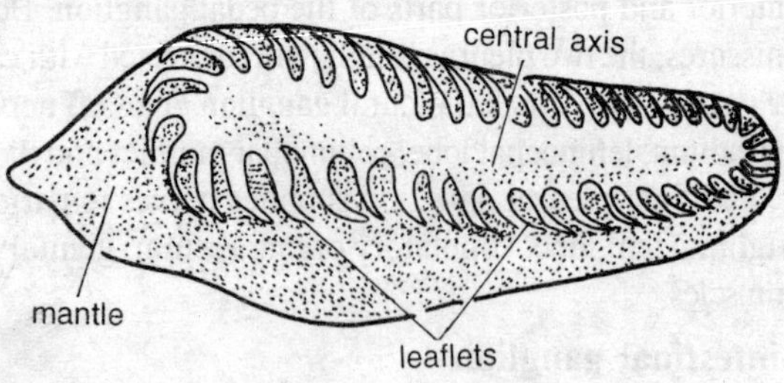

Fig. 5.21. *Pila*. Ospheridium.

In a transverse section, the osphradium consists of an outermost covering of a single layered epithelium, internally lined by a thin basement membrane, the interior is filled up with nerves, connective tissue and blood spaces. The epithelial cells are elongated possessing basal nuclei, and they are of three types (i) *sensory,* (ii) *ciliated* and (iii) *glandular*. The ciliated cells line the attached margin, while the sensory cells devoid of cilia cover the osphradium. The flask-shaped glandular cells are found scattered among the sensory cells.

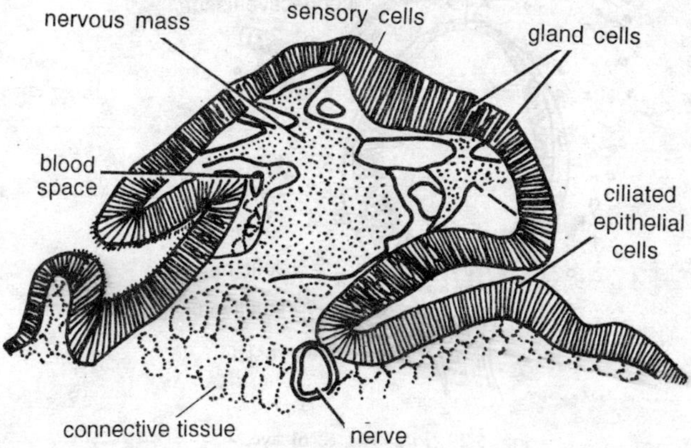

Fig. 5.22. *Pila*. T.S. of Ospheridium.

Formerly it was regarded as a rudimentary gill, but it has now been established that the osphradium serves to test the chemical and physical properties of water. The osphradium helps in the selection of food material (*Spengel*).

2. Eyes

Two eyes are borne on short stalks called *ommatophores* situated behind the second pair of tentacles.

In section, the eye is in the form of a pyriform cup or optic vesicle. The outer epithelium of the eye forming the *cornea* or *pellucida externa*, It consists of a row of squarish cells devoid of pigments. Lying below the cornea is thick layer of connective tissue made up of fibres and nuclei. This layer encloses the pyriform optic vesicles consisting of a transparent epithelium forming *inner cornea* or *pellucida interna* lying next to connective tissue. The ovoidal structureless lens occupies the central space of the vesicle. The sensitive part of the eye of the *retina* forms the outer coat of the vesicle made up of large cells filled with pigment granules. Retina consists of large pigmented cells, which are of two types : (1) large, broad cells with hair-like processes on their outer ends, the *visual cells* and (2) the compressed cells lying in between the visual cells, the *supporting* or *packing cells*. Both kinds of cells at the inner ends are connected with nerve-cells and branches of optic nerve, which enter the optic vesicle at its posterior end. Internally, the retina is bounded a basement membrane.

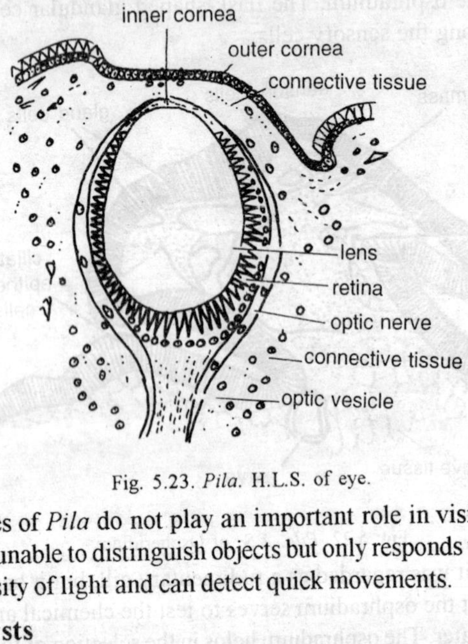

Fig. 5.23. *Pila*. H.L.S. of eye.

The eyes of *Pila* do not play an important role in vision. The snail seems to be unable to distinguish objects but only responds to the changes in the intensity of light and can detect quick movements.

3. Statocysts

The statocysts are two pyriform sacs of creamy white colour lying posterior to the pedal ganglion of each side and are not innervated from these ganglia. These are connected to cerebral ganglia by a pair of slender statocyst nerves. Each statocyst is filled with fluid and contains a single large *statolith* or few to many smaller bodies called *statoconia* floating in the fluid. Formerly it was supposed to be the organ of hearing and hence it was called *otocyst* but it is proved that these are the organs of equilibrium (*Delage*).

4. Tentacles

There are two pairs of *cephalic tentacles* in *Pila*, bordering the mouth. The anterior pair of tentacles is short, conical and contractile, called *labial palps* or *first pair of tentacles*. Behind them there is a pair of *True* or *second pair of tentacles*. They are tactile in function.

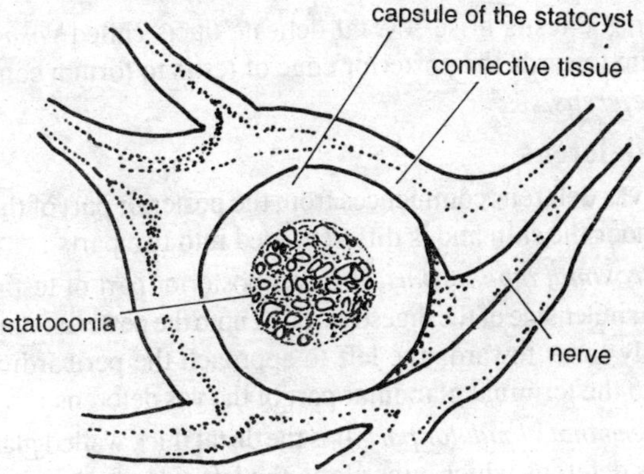

Fig. 5.24. *Pila.* Statocyst.

REPRODUCTIVE SYSTEM

Pila is *unisexual.* A well marked sexual dimorphism is seen. The shell of male is smaller than that of female. In male penis is well developed and the female has a more swollen body whorls than the male.

Male Reproductive Organs

The male reproductive organs comprise of the testis, vasa efferentia, vas deferens, seminal vesicle, penis and hypobranchial gland.

1. Testis

It is a single, flat, plate-like and roughly triangular whitish structure, occupying the upper part of the first two or three whorls of the shell. It lies in close association with the brownish or greenish digestive gland along its upper and inner or columellar side and appears quite distinct on account of its creamy colour. A thin cutaneous membrane separates the testis from the shell. The testis is made up of several seminiferous tubules lined by germinal epithelium.

The germinal epithelium of testis produces two types of sperms:

(i) *Eupyrene sperms.* These are small and thread-like. The body of each sperm consist of a head and a tail. These sperms are motile and fertilize the ova. They measure 25μ in length and 1.2μ in width.

(ii) *Oligopyrene sperms.* They are large semicircular or spindle-shaped. The head is large and the tail bears 4 to 5 cilia. They measure 30 to 40μ in length and 3μ in width. These are non-motile and incapable of fertilizing ova.

2. Vasa efferentia

From the testis arise several delicate ducts called *vasa efferentia*. They unite towards the posterior edge of testis to form a common duct, the *vas deferens*.

3. Vas deferens

The vas deferens commences from the posterior part of the testis and passes under the skin and is differentiated into two parts :

(i) *Proximal tubular part.* It forms posterior part of testis under the skin along inner side of the digestive gland upto the posterior renal chamber and finally turns towards the left to approach the pericardium. Here it opens into the terminal glandular part of the vas deferens.

(ii) *Terminal glandular part.* It is the distal thick walled glandular part of the vas deferens which runs along the left side of the rectum into the branchial chamber and opens into it by male genital pore. The male genital pore is situated a little behind the anus on a claw-like projection of genital papilla.

4. Seminal vesicle

It is situated on the right of the pericardium at the junction of the anterior and posterior renal chamber. It is swollen flask-shaped structure

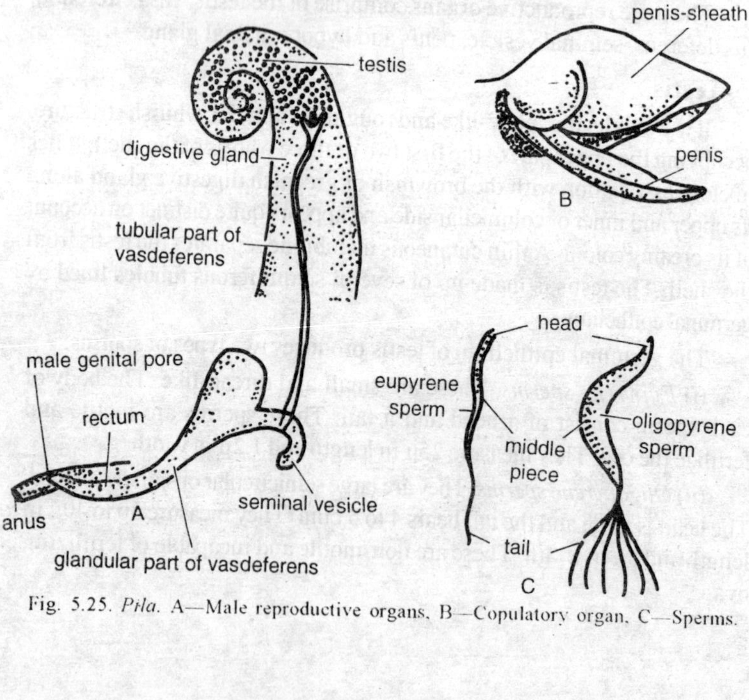

Fig. 5.25. *Pila*. A—Male reproductive organs. B—Copulatory organ. C—Sperms.

whose distal end is closed. The seminal vesicle stores the sperms and opens into the vas deferens at the junction of its tubular and glandular parts.

5. Penis or Copulatory organs

On the left side of the inner surface of the mantle, there is a thick glandular flap of yellowish colour. It is attached along its right side but its left free edge is rolled to form spout-like penis-sheath. Inside the sheath is the penis.

The penis is about 1.25 cm. long, provided with flagella arising from the roof of the mantle. It is slightly covered with swollen base and tapering end has a groove along its length on the inner side. It is capable of great extension.

6. Hypobranchial gland

It is a glandular structure with pleated surface at the base of penis-sheath. The gland is without duct and, therefore, its secretion is poured directly upon its surface.

Female Reproductive Organs

The female reproductive organs are ovary, oviduct, seminal recaptacle, uterus, vagina copulatory organ and hypobranchial gland.

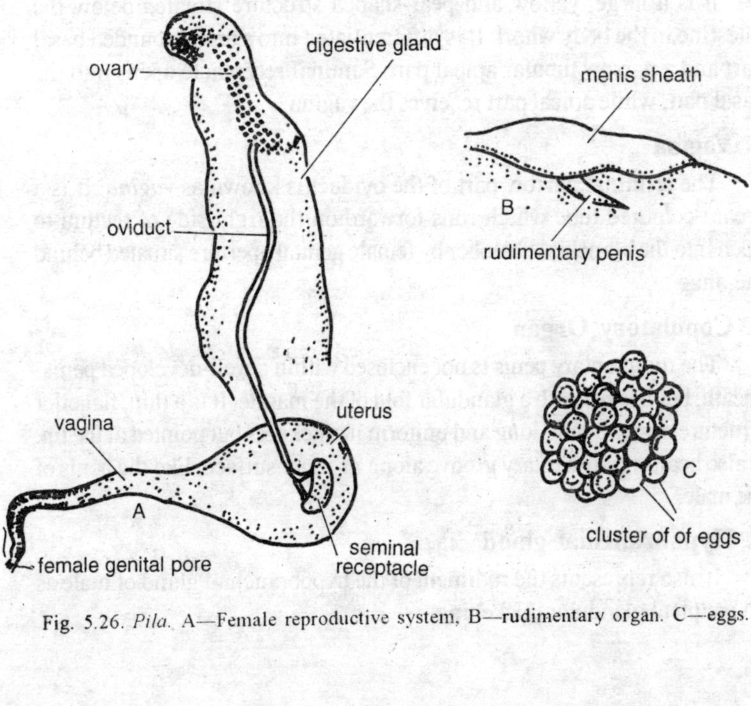

Fig. 5.26. *Pila*. A—Female reproductive system, B—rudimentary organ. C—eggs.

1. Ovary

The ovary is present in the same position as the testis in the male. It is found in two or three whorls of the shell. It is orange coloured structure remains attached to the inner surface of the digestive gland. Branches of the ovary are single layered, flask-shaped structures with slender tubular necks uniting together to form minute ducts which open into oviduct.

2. Oviduct

The oviduct is a semi-transparent, narrow tube arising from the middle of the ovary and passing downwards along the margin of the digestive gland. Finally, it ascends upwards to receive the opening of receptaculum seminis.

3. Seminal receptacle or Receptaculum seminis

The small, bean-shaped *receptaculum seminis* or *seminal receptacle* lies enclosed within the cavity of the posterior renal chamber, closely attached ventrally to a thin-walled pouch from the walls of the uterus, called the pouch of the receptaculum. It is meant for storing the sperms received from the male *Pila* during copulation.

4. Uterus

It is a large, yellow and pear-shaped structure situated below the intestine in the body whorl. It is differentiated into a broad rounded basal part and a narrow tubular apical part. Seminal receptacle opens into the basal part, while apical part receives the vagina.

5. Vagina

The terminal narrow part of the oviduct is known as *vagina*. It is a cream-coloured tube which runs forward on the right side of rectum to open into the branchial chamber by female genital aperture situated behind the anus.

6. Copulatory Organ

The rudimentary penis is not enclosed within a well-developed penis-sheath, but lies beneath a glandular fold of the mantle. It is a thin, flageller structure nearly 6 mm. long and uniform in thickness but pointed at the tip. It also bears a rudimentary groove along its inner surface, like the penis of the male.

7. Hypobranchial gland

It also represents the rudiment of the hypobranchial gland of male as an incipient glandular thickening.

Copulation

Snails breed during rainy season. Copulation takes place in the water of ponds or on their banks and continues for 3 hours or more. The penis becomes connected at the base with genital papilla, at the end of the terminal part of vas deferens, and along with the sheath is inserted into the mantle cavity of the female snail. The tip of the penis enters the vaginal opening and through this connection the spermatic fluid is transferred to the seminal receptacle of the female.

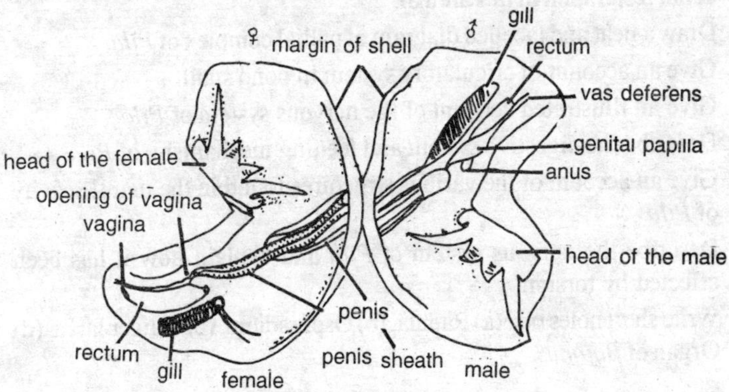

Fig. 5.27. *Pila.* Showing Copulation.

Fertilization

Fertilization is internal, but the development of the embryo takes place outside the body of female. Ovulation takes place two days after copulation. Female lays 200-280 eggs at a time in moist earth in a sheltered cavity near ponds and lakes.

Development

The eggs are rounded and as big as the pea-seeds. Each egg consists of an outer whitish shell, enclosing double shell membranes, a thick mass of solid white albumen and a little central mass of liquid albumen containing the embryo. During development the visceral mass and the shell of embryo become spirally coiled and a characteristic phenomenon of torsion takes place resulting in the asymmetry of the body. The young snails emerging from the eggs are similar to the adults.

Revision Questions

1. How is respiration effected in pond snail ? In what respects it differs from that of mussel ?
2. Describe the respiratory organs and mechanism of respiration in *Pila*.
3. Give an account of nervous system of *Pila*. Where would you place it in the scheme of classification.
4. Describe the reproductive organs of *Pila*.
5. Describe the respiratory organs of *Pila*. Describe the mechanism of aerial respiration in this animal.
6. Draw a neat and labelled diagram of pallial complex of *Pila*.
7. Give an account of circulatory system in pond snail.
8. Give an illustrated account of the nervous system of *Pila*
9. Describe the digestive system and feeding mechanism of *Pila*.
10. Give an account of the various structures found in the mantle cavity of *Pila*.
11. Describe the nervous system of *Pila* and explain how it has been affected by torsion.
12. Write short notes on : (a) Radula, (b) Osphradium, (c) Veliger larva, (d) Organ of *Bojanus*.

6
Unio

Unio comes under Class Pelecypoda or Bivalvia in which an external and internal symmetry is seen unlike gastropods. The cephalic region is rudimentary and the body is laterally compressed.

SYSTEMATIC POSITION

Phylum	-	Mollusca
Class	-	Pelecypoda
Order	-	Eulamellibranchiata
Genus	-	*Unio or Anodonta*

Distribution

The '*freshwater mussel*' is a familiar representative of Phylum Mollusca. The Family Unionidae to which *Unio* belongs, is cosmopolitan in distribution. The family comprises several genera and about 1000 species, of which a good number represented in India. *Anodonta* is the common representative of Britain, while in India *Unio* and *Lamellidens marginalis* are of common occurrence.

Habits and habitat

It is found in rivers, streams, lakes and ponds. During day time it usually lives partly burried in mud or sand at the bottom, keeping its posterior end exposed. It comes out to shallow places in night. They may also occur wedged in between the rocks and stones, with the valves slightly spread and the two siphons exposed. It slowly creeps with a plough share-like muscular foot. It hardly covers 16 kilometres in a year. Sexes are separate but there is no sexual dimorphism externally. Development is indirect through a larval stage and takes place in brood chamber, which are enlarged water tubes of female's gills.

External features

Shape and size

The body is soft, unsegmented, bilaterally symmetrical and laterally compressed. It is elongated and enclosed, and protected in a bivalve shell. The size varies from 2″ to 4″ in length.

Shell

A typical bivalve shell consists of two similar more or less oval, usually convex valves which are attached and articulated dorsally with each other. Each valve bears a dorsal protuberance called the *umbo* which rises above the line of articulation. The umbo is the oldest part of the shell, and the concentric lines around it are lines of shell growth. The two valves are attached by an elastic protein band called the *hinge ligament*, which is covered above with the *periotracum*. The hinge ligament is so constructed that when the valves are closed, the dorsal part is stretched and the ventral or inner part is compressed. Thus, when adductor muscles relax, the ligament causes the valves to open. The inner surface of each valve possesses, dorsally along the hinge line, small sharp ridges and teeth-like projections separated by grooves or sockets. These are called *hinge teeth*. The teeth of one valve fit into the corresponding sockets of the other to prevent lateral slipping. However, a dorsal fold of mantle tissue still lies between the tooth and socket of two opposing valves.

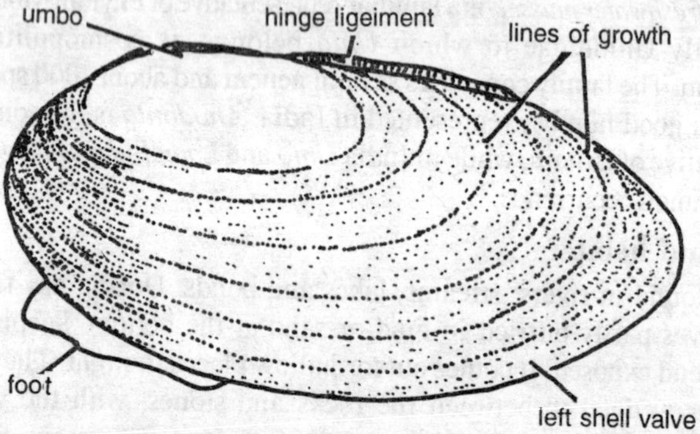

Fig. 6.1. *Unio*. External feature

The valves of the shell are pulled together by two large dorsal muscles, called *adductors*. The posterior muscle being larger than that of the anterior. Near the impression of the anterior adductor muscle are two small impressions, the dorsal and posterior one left by *anterior retractor muscle* and a ventral and posterior by *protractor muscle*. A small impression of a *posterior retractor muscle* also lies dorsal to the impression of posterior adductor muscle.

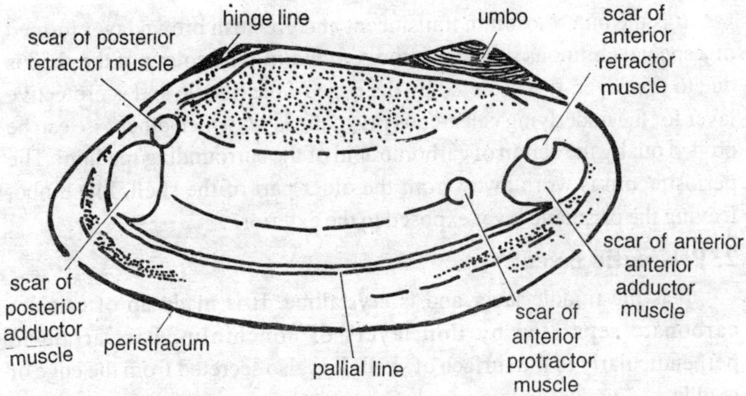

Fig. 6.2. *Unio*. Inner view of left shell.

Running parallel to the ventral margin of the valves from one adductor impression to the other is a thick line, known as *pallial line,* which is caused due to the insertion of muscle fibres from the edges of the mantle.

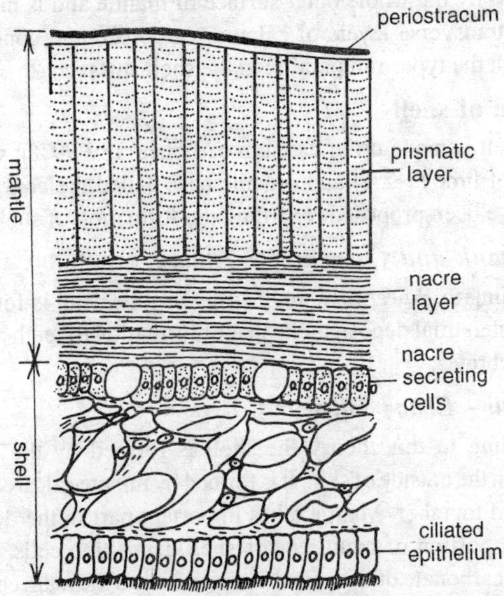

Fig. 6.3. *Unio*. T.S. of shell and mantle.

MICROSCOPIC STRUCTURE OF SHELL

The shell consists of three distinct layers

1. Periostracum

It is the outermost thin translucent and greenish brown layer formed of a special chitinous substance, the *conchiolin*. The colour of the shell is due to this layer. It is secreted by the edge of the mantle and is protective layer for the underlying calcareous prismatic layer which otherwise can be eroded out by the action of carbonic acid of the surrounding medium. The periostracum is worn away from the older part of the shell, like umbo, leaving the prismatic layer exposed to the exterior.

2. Prismatic layer

It is the middle layer and is crystalline. It is made up of calcium carbonate separated by thin layers of conchiolin. It is arranged perpendicularly to the surface of shell. It is also secreted from the edge of mantle.

3. Nacreous layer or pearly layer

It is the innermost layer of shell and also known as the *mother of pearl* or *nacre,* since it presents a smooth, iridescent or lusterous surface. It is secreted by the whole outer surface of mantle and is in the form of alternative transverse layers of calcium carbonate and conchiolin. The layer of both the types run parallel to the shell surface.

Formation of shell

The shell is made up of calcium carbonate (89-99%), very little of phosphate of lime (1-2%) and organic base called the conchiolin. Two theories have been proposed regarding the formation of shell.

1. Bowerbank and Carpenter's Theory

According to *Bowerbank* and *Carpenter*, the shell is formed by the growth of interstitial deposits, like that of formation of teeth and bones in higher vertebrates.

2. Reaumur - Eisioz Theory

According to this theory the shell is formed by the cuticle-like deposition on the outside of skin. It is formed by minute calcareous particles which bound together. Mantle takes important part in the deposition of shell. Every portion of mantle bears shell deposition cells. These cells separate the carbonate of lime from blood, which is the main constituent of shell. The umbo portion of shell is secreted first and subsequent layers are added in gradual succession forming the lines of growth.

The edge of the mantle bears three folds an inner, middle and outer. The outer fold is related to secretion of the shell. The inner surface of

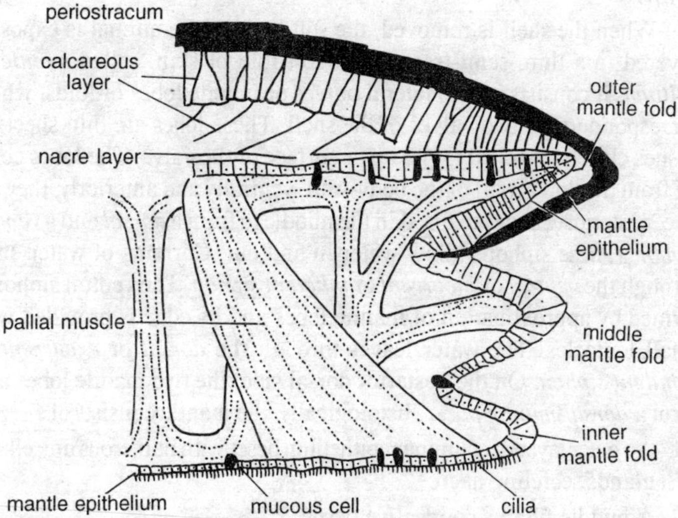

Fig. 6.4. *Unio.* T.S. through margin of shell.

outer fold lays down the periostracum, and the outer surface secretes the first calcareous layer. The entire mantle surface secretes the remaining calcareous shell.

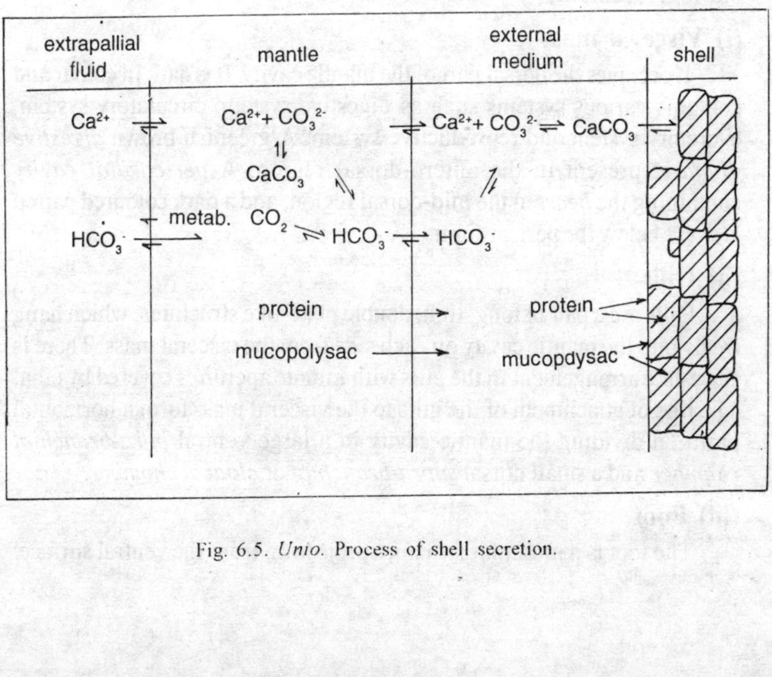

Fig. 6.5. *Unio.* Process of shell secretion.

Mantle

When the shell is removed, the soft body of the animal is exposed, covered in a thin, semi-transparent soft fold of skin, called *mantle* or *pallium*. It consists of two lateral halves or mantle lobes or folds, which correspond to the two valves of the shell. These lobes are thin sheets of tissues closely attached to the inner surface of the valve. The lobes come off from the dorsal part of the body-wall, ventrally and anteriorly, they are free, posteriorly, they are fused in the middle to from a *dorsal* and a *ventral siphon*. These siphons allow water in and out. Currents of water enter through the *ventral* or *incurrent* or *inhalent siphon*. The ventral siphon is formed by approximation of mantle lobes and its edge is provided with small tentacles. The water leaves through the *dorsal* or *excurrent* or *exhalent siphon*. On the posterior dorsal side, the two mantle lobes also form a *dorsal-mantle pore*. Histologically, the mantle consists of :

1. An outer layer of columnar epithelium beset with numerous unicellular glands secreting nacre.
2. A middle fibrous connective tissue.
3. An inner ciliated epithelium containing mucous secreting cells.

MANTLE CAVITY OR PALLIAL CAVITY

The space enclosed between the two lobes of mantle is called *mantle cavity*. It encloses *viseral mass, gills, foot* etc. which are exposed when one fold is removed.

(i) Visceral mass

It occupies the dorsal part of the mantle cavity. It is dark in colour and contains various systems such as digestive system, circulatory system, excretory system and reproductive system. A greenish-brown *digestive gland* is present in the antero-dorsal region. A *pericardial cavity* containing the *heart* in the mid-dorsal region, and a dark coloured paired *kidneys* below the pericardium.

(ii) Gills

Gills are a pair of long, thin, double plate-like structures, which hang freely into the mantle cavity on each side from the visceral mass. There is sieve-like arrangement in the gills with minute apertures covered by cilia. The line of attachment of the gills to the visceral mass form a horizontal partition dividing the mantle cavity in a large ventral *infra-branchial chamber* and a small dorsal *supra-branchial* or *cloacal chamber*.

(iii) Foot

The foot is a muscular structure projecting from the ventral surface.

Its size and form are variable depending upon the life of the animal. The mass of the foot is invaded by a portion of viscera, at least the alimentary canal, liver superficially and the gonads in the life of animal.

(iv) Head

A distinct head is absent in *Unio*. The eyes and tentacles are also absent. The large mouth opens beneath the anterior adductor muscle. It is boarded by a pair of broad, lamellar *labial palps* on either side.

Locomotion

The locomotion is performed by muscular foot. It is affected by a combination of blood pressure and muscle action. Engorgement of the foot with blood, coupled with the action of pair of *pedal protractor muscles,* produces extension. The water in mantle cavity and the blood act as a hydrostatic skeleton. The pressure of these two fluids is elevated by the adducted valves. Blood from the visceral mass is forced down into the pedal haemocoel, causing the foot to dilate and anchor into the substratum.

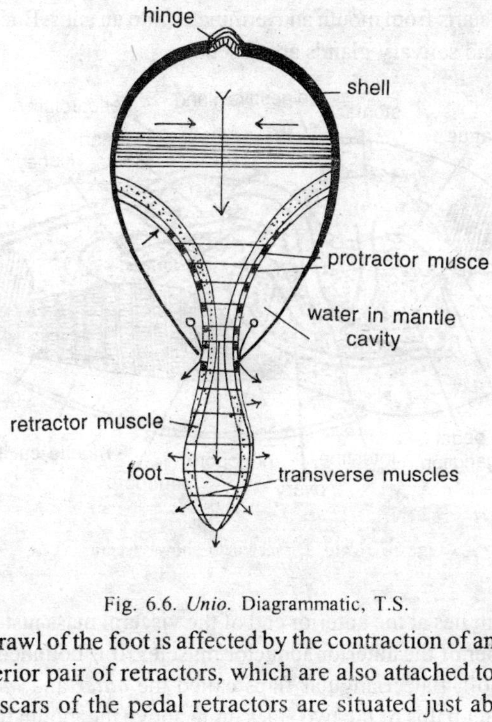

Fig. 6.6. *Unio*. Diagrammatic, T.S.

Withdrawl of the foot is affected by the contraction of an anterior pair and a posterior pair of retractors, which are also attached to the foot and shell. The scars of the pedal retractors are situated just above those of adductors.

Coelom

Coelom is greatly reduced due to the development of connective tissue, unstriated muscle-fibres and blood-sinuses. Coelom is a *haemocoel*, filled with blood. It is represented by three small cavities : (a) a single ovoidal chamber, called *pericardium*, which is situated dorsally and in it, heart is located, (b) the cavity of the kidney or *urocoel*, and (c) *gonocoel* or cavity of gonad. Originally, all the three were intercommunicated but due to gradual separation, the cavities of the gonads are completely separated, while the cavities of kidney and pericardium are inter-communicated.

DIGESTIVE SYSTEM

The digestive system comprises of alimentary canal and a paired digestive glands.

I. Alimentary canal

The alimentary canal is complete. It is a coiled tube of varying diameters and starts from mouth and terminates into an anus. Buccal cavity, jaws, radula and salivary glands are absent.

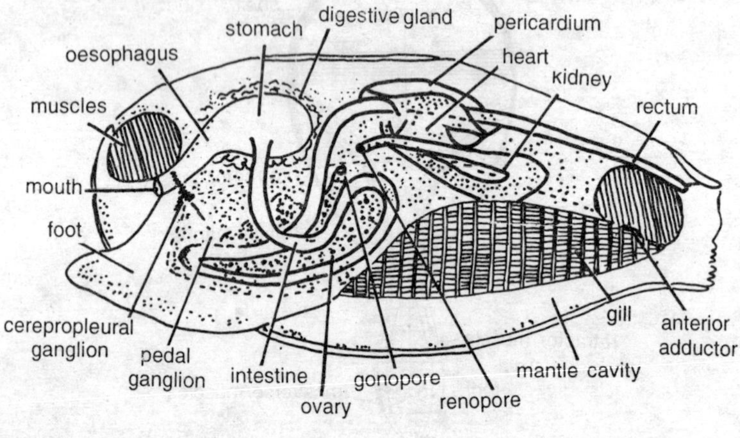

Fig. 6.7. *Unio.* Dissected to show viscera.

1. Mouth

The mouth lies at the anterior end of the visceral mass just below the posterior border of the anterior adductor muscles. It is bounded on either side by two soft, flat, triangular flaps called the *outer* and *inner labial palps.* The outer palps of the two sides unite above the mouth to from the *upper lip.* The inner palps of the two sides likewise fuse below the mouth to form *lower lip.* The outer surface of the palps is heavily ciliated. The mouth leads into the oesophagus.

2. Oesophagus

The oesophagus is a short, wide ciliated tube, which leads into stomach. The radula is absent as it is useless for the animal since it feeds on minute organisms.

3. Stomach

The stomach is a wide thick-walled rounded sac, situated dorsally in the visceral mass and is surrounded by paired digestive glands. The stomach is produced into a pouch like out-growth known as *pyloric caecum.* It contains a transparent gelatinous flexible rod called the *crystalline style*. It has proteinaceous matrix deposited with amlyase which is a carbohydarate splitting enzyme. The style rotates due to cilia of the stomach and brings about the mixing of enzymes with food. The crystalline style is secreted by the cells of the stomach and is contained in a *style sac*.

4. Intestine

From the posterior end of stomach arises the intestine and enters the substance of foot where it coils on itself and again ascent the place where the pericardial cavity starts. It then runs dorsally again and is continued into the rectum.

5. Rectum

It is the terminal part of intestine which starts from pericardium and runs for a short distance before opening through the anus in the cloacal chamber. Internally the rectum has a ridge, like the *typhlosole* of earthworm.

6. Anus

The anus is situated on an anal papilla, just above the posterior adductor muscle and opens in the cloacal chamber.

II. Digestive glands

The digestive glands are a pair of large irregular green or dark-brown masses, situated on the sides of the stomach. They lead into anterior part of stomach by many ducts. They are made up of highly branched tubules, line by columnar epithelium of unequal height containing basal nuclei. The cells of these glands also swallow the food particles and break the solid particles, in addition to the secretion of enzymes. These cells intracellularly digest the protein and fats, and absorb the carbohydrates.

Food and feeding

Freshwater mussel is a ciliary feeder. If feeds upon micro-organisms and organic debris which are filtered from the respiratory nutritive current of water drawn into mantle cavity. A continuous current of water is maintained mainly by the action of lateral cilia of gill filaments. As the water current enters the mantle cavity through the inhalent siphon it slows down and the solid food particles settle down over the gill surface. These are removed to the edge of mantle where they are pushed back and are expelled out. The lighter food particles get entangled into the mucus secreted by the gills. By the beating of the cilia of labial palps, the mucus is drwan towards mouth. The particles of food which cannot be digested are dropped down into the mantle cavity, whereas the digestible particles are pushed into the mantle cavity, along the deep ciliated grooves between the labial palps.

Digestion

The digestion is *intracellular.* Food from the stomach enters the digestive glands, the cells of which readily ingest and break down solid particles. Protein and fat digestion is exclusively intracellular, and the cells of glands also absorb carbohydrates. The only extracellular enzyme is the carbohydrates digesting enzyme (*amylase*) released in the stomach by the dissolution of the crystelline style. The style pouch is lined with cilia, the beating of which rotates the style and moves it forward with the result that its free end constantly rubs against a particular portion of the stomach wall and its head is worn away getting mixed with stomach content. Digested food diffuses from the glands into the blood for distribution.

Undigested food particles are sent back into the stomach, whence they pass into the intestine and then into the rectum by ciliary action. Function of intestine or rectum is not properly understood. They may absorb water from faecal matters. From the rectum, faeces is passed through the anus into the cloaca, which it leaves with the outgoing water current through exhalent siphon.

RESPIRATORY SYSTEM

The respiratory organs in *Unio* are a pair of *gills* or *ctenidia* and *mantle*.

Ctenidia or gills

There are two gills that hang in the mantle cavity, one on either side between the mantle and visceral mass. The gills are plate-like hence they are called *lamellibranches.* Each gill has a central axis and plate like structure on either side. Hence the gills are called *bipectinate gills.*

Each gill is a W-shaped structure. It consists of two plate-like structures called *laminae.* One lamina is situated on the inner side called the *inner*

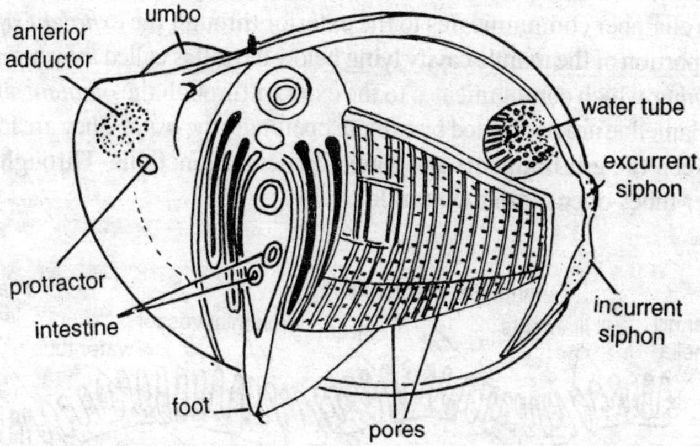

Fig. 6.8. *Unio*. A schematic section of the mussel showing the respiratory organs. Water tubes have been exposed, whereas, the pores are highly magnified.

lamina and the other is situated on outer side and is called *outer lamina*. Each lamina is V-shaped structure and formed of two vertical plates called *lamellae*. The outer and inner lamellae are united together at their anterior, ventral and posterior edges, but are free dorsally. Each lamella is formed of a series of vertical filaments called *gill filaments*. The gill filaments of the two lamellae of a lamina are continuous at the free ventral side. The gill filaments are joined to one another by horizontal bars called *inter-filamentar junctions*. Two lamellae of a lamina are joined together by *inter-lamellar junctions*. The inter-lamellar junctions divide the cavity lying between the lamellae into vertical chambers called *water tubes*. Dorsally

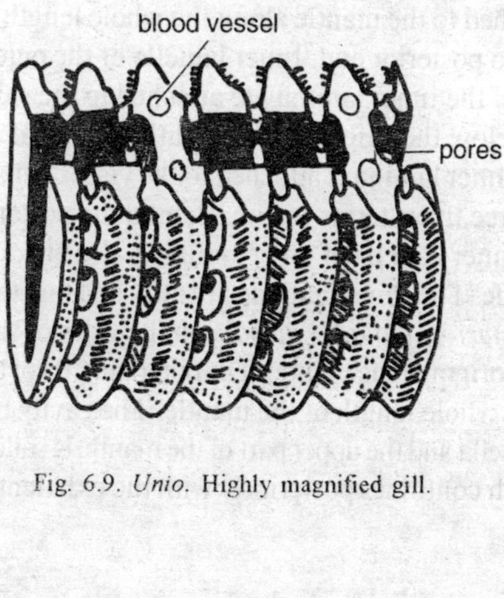

Fig. 6.9. *Unio*. Highly magnified gill.

all the water tubes open into a large cavity called *supra-branchial chamber.* This chamber communicates to the exterior through the *exhalent siphon.* The portion of the mantle cavity lying below the gill is called *infrabranchial chamber* which communicates to the exterior through the *inhalent siphon*. The lamellae are perforated by minute openings, the *ostia.* They are located between the gill filaments and inter filamentar junctions. Through ostia water tubes opens into the mantle cavity.

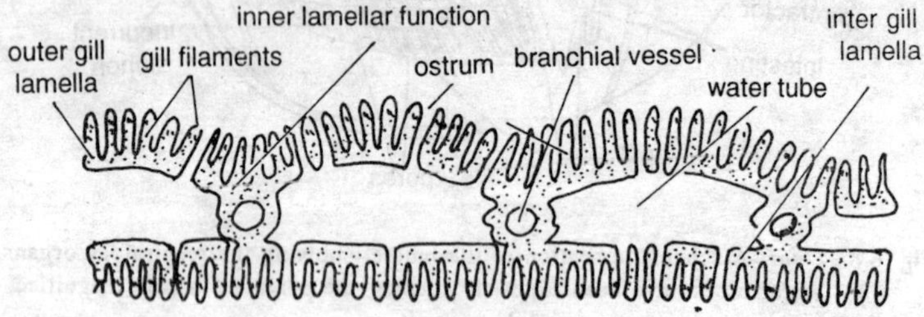

Fig. 6.10. *Unio.* T.S. of outer gill lamella.

Each gill filament is covered by a layer of ciliated epithelial cells. The cilia are of three types, those present an outer face of the filaments are called *frontal cilia*, those on lateral parts are *lateral cilia* and those lying in between are *letero-frontal cilia.* Internally the gill filament is supported by chitinous rods.

Attachment of gills

The mode of the attachment of the gills is very peculiar, as it determines the course of the water-current in the body. The outer lamella of the outer lamina is attached to the mantle along the whole length by its upper edge from anterior to posterior end. Inner lamella of the outer lamina and the outer lamella of the inner lamina are attached to the sides of the visceral mass a little below the origin of the mantle by dorsal edges. The inner lamella of the inner lamina is attached to the visceral mass in the anterior region, but is free in posterior region. Behind the posterior margin of the visceral mass inner lamella of inner lamina is also attached to its fellow of the opposite side. The inner laminae are attached together by their dorsal edge to the *posterior adductor muscle* on the sides of the visceral mass a little below the origin of the mantle. The outer lamella of the outer lamina is attached to the whole length of the mantle. The cavity between the upper edge of gill-lamella and the upper part of the mantle is called supra-branchial chamber, which continues posteriorly with the exhalent siphon.

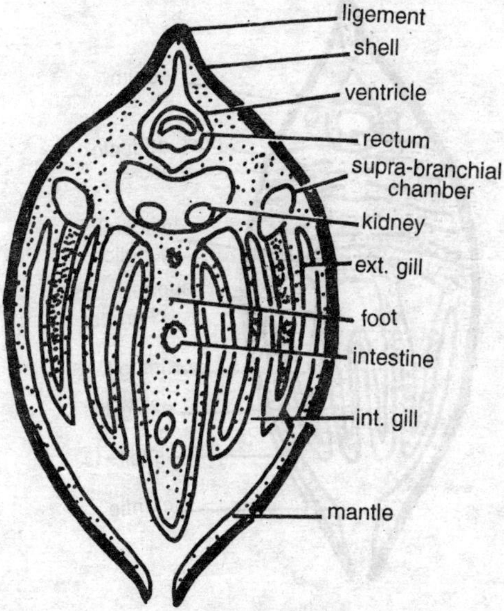

Fig. 6.11. T.S. of *Unio* through anterior region.

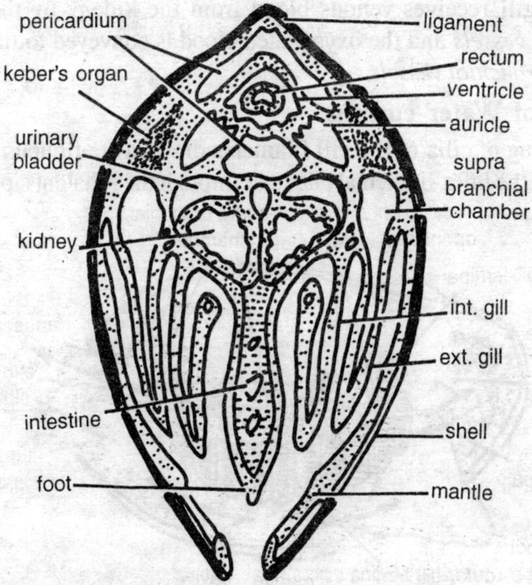

Fig. 6.12. T.S. of *Unio* through middle region.

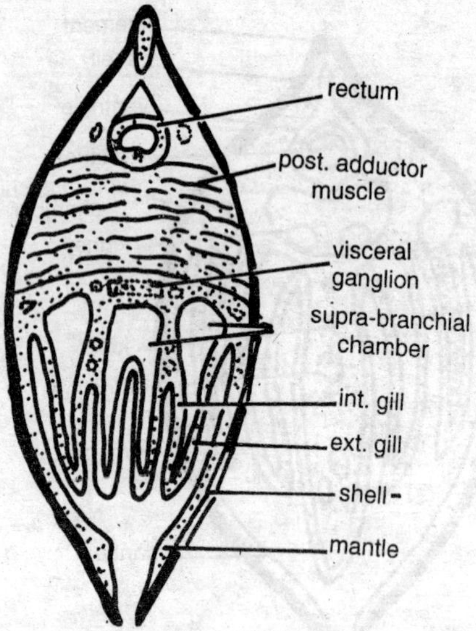

Fig. 6.13. T.S. of *Unio* through posterior region.

Blood supply

The gill receives venous blood from the kideny by the *afferent branchial vessels* and the oxygenated blood is conveyed to the heart by *efferent branchial vessels.*

Course of Water current

Beating of cilia of the gill filaments draws a continuous current of water into the infra-branchial chamber through the inhalent siphon. From

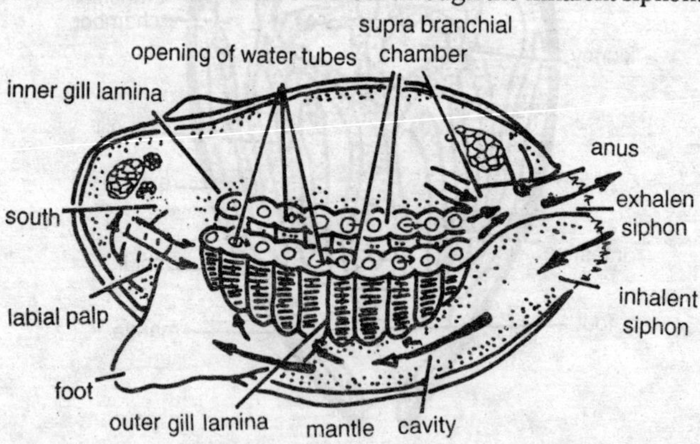

Fig. 6.14. *Unio*. Course of water current

here it enters the gills through ostia. Inside the gill, the water fills the watertubes and flows out into the supra-branchial chamber placed dorsal to each gill lamina and open posteriorly into the cloacal chamber. The water finally leaves the cloacal chamber through exhalent siphon which discharges water to the exterior.

Supra-branchial chamber ⟶ Cloacal chamber
↑ ↓
Water tubes (inside gills) Exhalent siphon
↑ ↓
Ostia ⟵ Sub-branchial or infra-branchial chamber ⟵ Inhalent siphon ⟵ Surrounding water

CIRCULATORY SYSTEM

The circulatory system of *Unio* is said to be *open type* because the capillaries are absent. The circulatory system comprises three parts.

1. Heart
2. Blood vessels
3. Blood

1. Heart

The heart is situated dorsally near the posterior adductor muscle. It is enclosed in *pericardium.* The pericardium is a thin walled triangular chamber. It is lined by epithelium and is filled with a fluid and represents a part of coelom. It is communicated with the supra-branchial chamber through the kidneys.

The heart is three chambered structure, having two *auricles* or *atria* and a *ventricle.* The auricles are thin-walled, highly distensible, triangular chambers, one on either side of ventricle. They are attached by their broad bases to the pericardium and open dorsally into the ventricle through *auriculo-ventricular apertures.* Each auriculo-ventricular aperture is provided with a valve that opens towards the ventricle and prevents the backward flow of blood. The ventricle has the form of a horizontal tube and surrounds a part of rectum.

The heart consists of smooth muscle-fibres covered externally by an epithelium called the *epicardium.* Endocardium is absent. The heart beats 20-100 times per minutes.

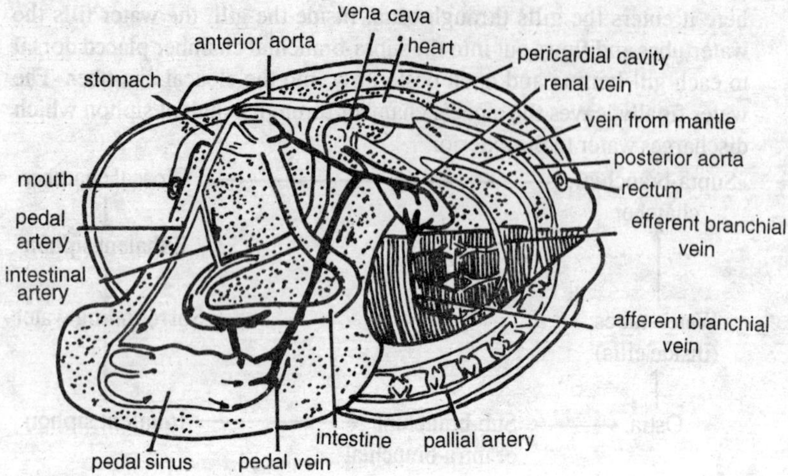

Fig. 6.15. *Unio.* Circulatory System.

2. Blood vessels

The blood vessels comprise the :

(i) arteries

(ii) sinuses

(iii) veins,

(i) Arteries

From each end of ventricle arise, an *anterior aorta* passing above and a *posterior aorta* below the rectum. The anterior aorta supplies to the anterior parts of the body i.e. mantle, foot and visceral mass by *anterior pallial, pedal* and *visceral arteries*. The posterior aorta supplies to the posterior parts of the body i.e. rectum, mantle, nephidia, pericardium etc. by *rectal, posterior pallial, nephridial,* and *pericardial arteries* respectively.

(ii) Sinuses

The arteries ramify all the organs and finally empty into sinuses. From small sinuses blood is sent to larger sinues. The sinuses lack the epithelial lining of true blood vessels and connect directly with veins.

(iii) Veins

Blood from different parts of body is collected by many small veins, and at the end it reaches the large vein, called *vena cava* which is placed longitudinally in the body. The vena cava collects blood from kidneys by the *renal vein,* here nitrogenous wastes are removed. Blood from kidneys is sent to the gills for aeration by *afferent branchial veins,* which gives off branches to all the gill-filaments. The oxygenated blood is sent to the auricles by *efferent branchaial veins.* Some deoxygenated blood is also sent directly to the auricles by vena cava. Blood from each mantle lobe is collected by a *pallial vein,* which also opens into the auricle of its side.

3. Blood

The blood consists of *plasma* and *corpuscles.* The plasma is colourless in some species where as faint blue in others due to presence of *haemocyanin,* a respiratory pigment. The corpuscles are colourless and amoeboid hence called the *leucocytes.* Some are granular, others are non-granular. Blood helps in transportation of food, oxygen, carbon dioxide and waste products. Corpuscles also help in phagocytosis of microorganisms.

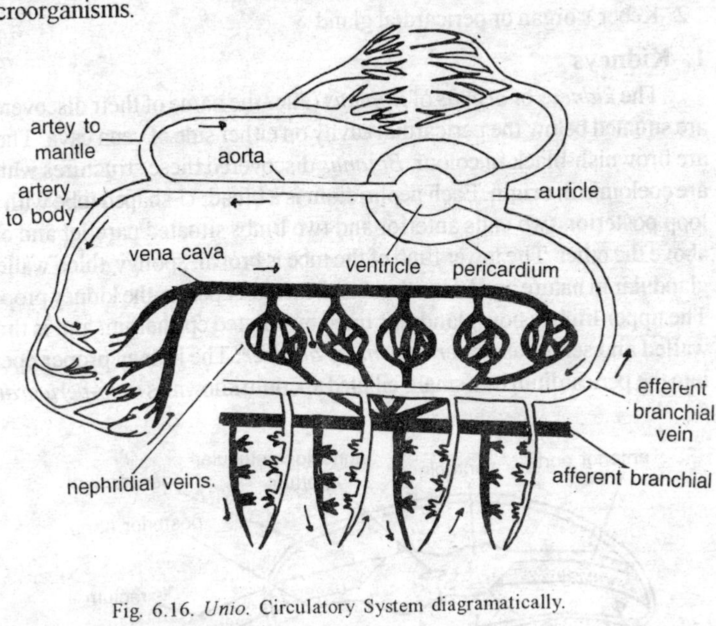

Fig. 6.16. *Unio.* Circulatory System diagramatically.

Course of circulation

The auricles receive the purified or oxygenated blood and the ventricle pumps it to the various body organs in both forward and backward directions. The various veins and sinuses collect blood from body organs and carry it to the kidneys and gills for purification or aeration. The course of blood is again carried to the auricles. The course of blood circulation can be shown as under :

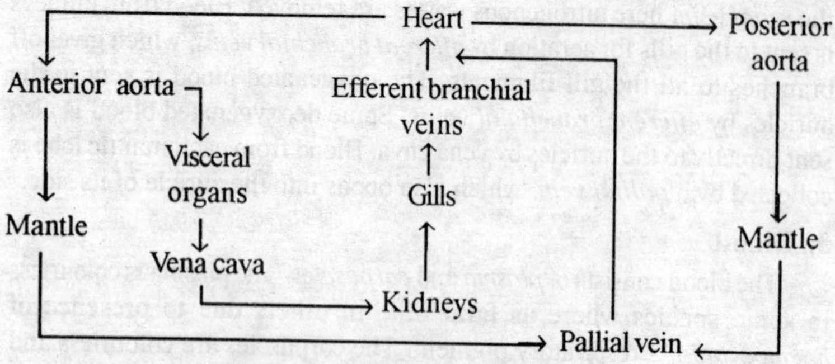

EXCRETORY SYSTEM

In *Unio* followings are the excretory organs.

1. A pair of kidneys or organs of Bojanus
2. Keber's organ or pericardial gland

1. Kidneys

The *kidneys* or *orgnas* of *Bojanus* (after the name of their discoverer) are situated below the pericardial cavity on either side of vena cava. These are brownish-black in colour. *Bojanus* discovered these structures which are coelomic in origin. Each nephridium is a broad, U-shaped tube with its loop posterior, two ends anterior and two limbs situated parellel and one above the other. The lower limb of the tube is brown, spongy, thick walled, glandular in nature and forms the actual excretory part or the kidney proper. The upper limb is non-glandular lined by ciliated epithelium and is thin-walled and serves as *ureter* or *urinary bladder*. The kidney proper opens into the pericardium by a small, ciliated aperture known as *reno-pericardial*

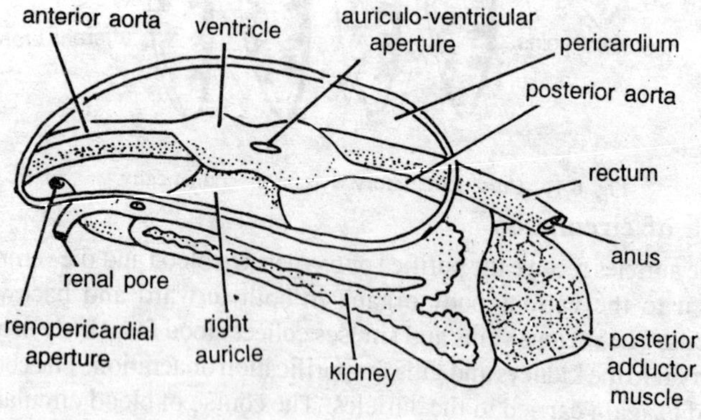

Fig. 6.17. *Unio*. Excretory System.

aperture. The ureter opens into the supra-branchial cavity by a small renal aperture between inner gill lamina and visceral mass.

The ventral glandular part of the kidney extracts nitrogenous products and guanin from the pericardial fluid and the blood. The ciliated epithelial lining of the bladder creates an outgoing current, which carries excretory fluid to the outside.

2. Keber's Organ

Keber's organ is a large reddish-brown gland situated in front of pericardium. It discharges waste products in the pericardium, from where it is eleminated out side via kidneys.

NERVOUS SYSTEM

Due to sluggish, sedentary life, the nervous system of *Unio* is greatly reduced. It consists of three pairs of ganglia, connectives and commissures. These are :

1. Cerebro-pleural ganglia

At the base of the labial palps, just outside the corners of the mouth on either side, is placed a small yellowish, somewhat triangular *cerebropleural ganglion.* Each is formed by the fusion of two ganglia : a cerebral and a pleural. The ganglia of both the sides are equivalent of the brain. They are connected with each other by a thin transverse *cerebral commissure*, which passes over the oesophagus. Each cerebro-pleural ganglion gives out-(i) an *anterior adductor nerve* to the corresponding muscle, (ii) a *labial nerve* to the labial palp, and (iii) an *an anterior pallial nerve* to the anterior part of the mantle. Besides, each ganglion gives off

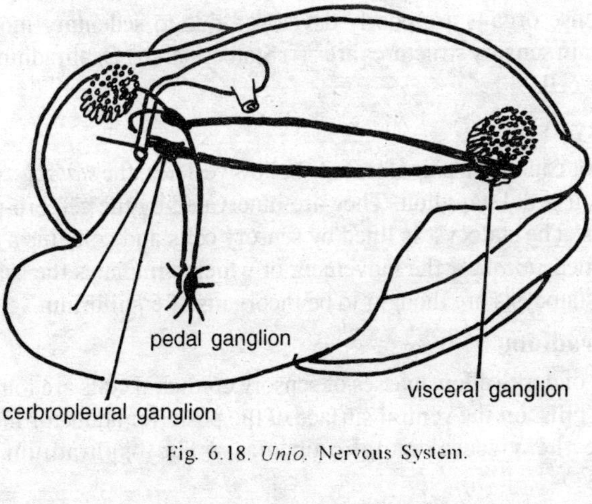

Fig. 6.18. *Unio.* Nervous System.

two conspicuous connectives, a *cerebro-visceral connective* which runs posteriorly to unite with the *viceral ganglion* of that side, and a *cerebro-pedal connective* which passes ventrally into the foot to unite with the *pedal ganglion* of the same side.

2. Pedal ganglia

The pedal ganglia are situated in the foot about 1/3 from its anterior end. They are present above muscular part and below the visceral mass. Both the pedal ganglia are closely associated so as to form a bilobed mass which supplies the nerves to the foot, its muscles and the statocysts. Each pedal ganglion is connected to the cerebro-pleural ganglion of its side by the cerebro-pedal connective.

3. Visceral ganglia

A pair of visceral ganglia are situated mid-ventrally on the posterior adductor muscle. The visceral ganglia are fused together to form a flattened, somewhat rectangular mass. On each side, the visceral ganglionic mass gives off (i) *a dorsal pallial nerve*, (ii) *a posterior pallial nerve* to the posterior part of the mantle, (iii) a *posterior renal nerve* to the kidney, (iv) *branchial nerve* to the gill. It is connected with the cerebro-pleural ganglion of its side by a thin cerebro-visceral connective which gives off several small nerves to the visceral mass.

Autonomous nervous system

Recently, *E.I.Thomas* (1974) has worked out anatomy of *Lamellidens* and has reported the presence of minor ganglia in the chief muscles and has suggested that they are autonomous.

SENSE ORGANS

The sense organs are poorly developed due to sedentary mode of life. The main sensory structures are : (i) Statocysts, (ii) Osphradium and (ii) Sensory cells.

(i) Statocysts

The foot contains a pair of minute hollow vesicles, the *statocysts,* one close to each pedal ganglion. They are innervated by the cerebro-pedal connectives. The statocyst is lined by sensory cells and contains a mass of lime, called *statolith,* the movement of which stimulates the sensory cells. The statocysts are thought to be theorgans of equilibrium.

(ii) Osphradium

A pair of dark yellow masses of sensory epithelial cells are found at the base of gills, on the ventral surface of the posterior adductor muscle and above the visceral ganglionic mass. The osphradium is a *chaemoreceptor* as it tests the chemical nature of water entering the mantle cavity by way of the inhalent siphon.

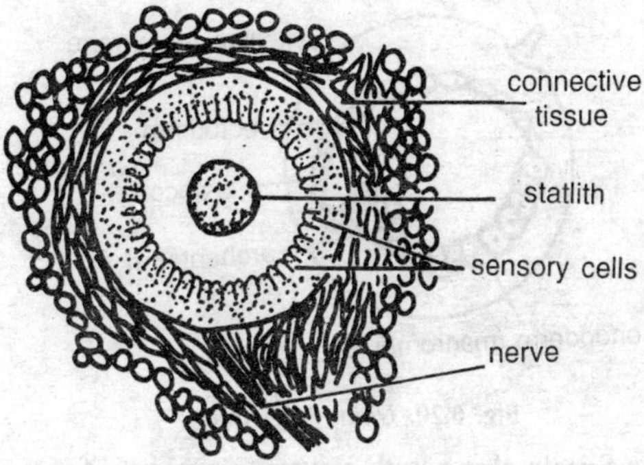

Fig. 6.19. *Unio.* Statocyst.

(iii) Sensory cells

There are also other light responding devices found along the edges of mantle lobe and abundant on the inhalent siphon. They are regarded to be tactile organs and are also sensitive to light.

REPRODUCTIVE SYSTEM

The *Unio* is *dioecious* i.e. sexes are separate. The reproductive organs consist of the gonads and their ducts.

Male reproductive organs

These consist of a pair of *testes* which are white-coloured, highly branched, glandular masses situated in the visceral mass between the loops of intestine. From each testis a *vas deferens* arises which opens into the supra-branchial chamber through *male genital pore* situated in front of renal aperture.

Female reproductive organs

Ovaries are a pair of pink coloured glandular mass of branched tubules occupying the same position as testes in male. These become greatly enlarged during breeding season. Each ovary gives out an *oviduct* which opens through *female genital aperture* situated in front of renal opening. The accessory reproductive organs are absent.

Fertilization

The eggs are fertilized in the outer gill laminae. The eggs are collected in the supra-branchial chamber. From here they reach the water tubes of the outer gill-laminae and are held there by the mucous. The sperms come

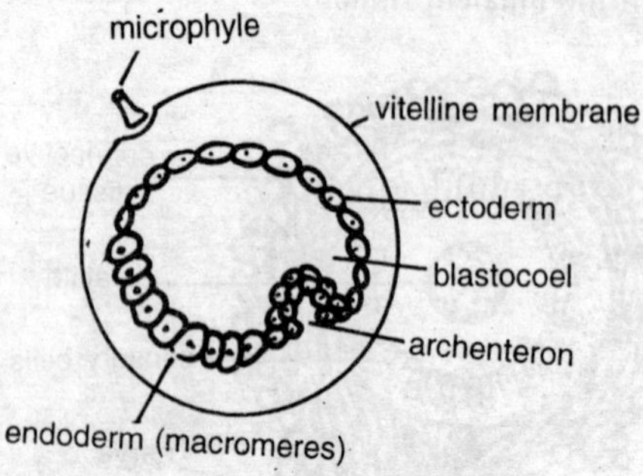

Fig. 6.20. *Unio*. A gastrula

out of the body of male along with outgoing current of water. Some of these reach the gill laminae of the female through the incurrent siphon with water current and fertilize the eggs.

Development

The fertilized ova or zygotes undergo segmentation within the outer gill laminae. The cleavage is complete (*holoblasitic*) but unequal. A hollow, spherical embryo, the *blastula* with a fluid filled cavity, the *blastocoel,* is formed. The bigger blastomeres are called *macromeres* and the smaller ones, *micromeres*. During gastrulation some of the macromeres are invaginated, resulting in the formation of an exceedingly small cavity, the *archenteron*. Later on there is an extensive invagination of macromeres which very greatly reduces the blastocoelic cavity. This is the rudiment of *shell gland*. Some large cells are budded off into the blastocoel, where they aggregate to form *mesoderm*. The invagination of the macromeres which constitute the rudiment of the shell gland, marks the dorsal surface of embryo, while the posterior end is marked by a tuft of cilia. The cells of the shell gland, which are tall, columnar, secrete an unpaired shell. Later arises a shell consisting of two valves, which are more or less triangular in appearence and have their ventral ends sharply incurved to form *hooks* that bear *spines*. The mass of cells consisting the embryo is divided into two on the ventral side, but remains undivided dorsally. The two ventral halves are the rudiments of mantle lobes. These lobes bear peculiar brush-like sense organs. A long, thread-like adhesive organ, the *provisional byssus,* appears on the ventral surface of the body. The mesoderm produces a large *adductor muscle* that stretches between the two valves of the shell. Gradually a *glochidium larva* is formed. The glochidium is the characteristic of freshwater mussels. Within the gill lamina, which serves as a *brood pouch*, the larva, is nourished by the secretions of laminar wall

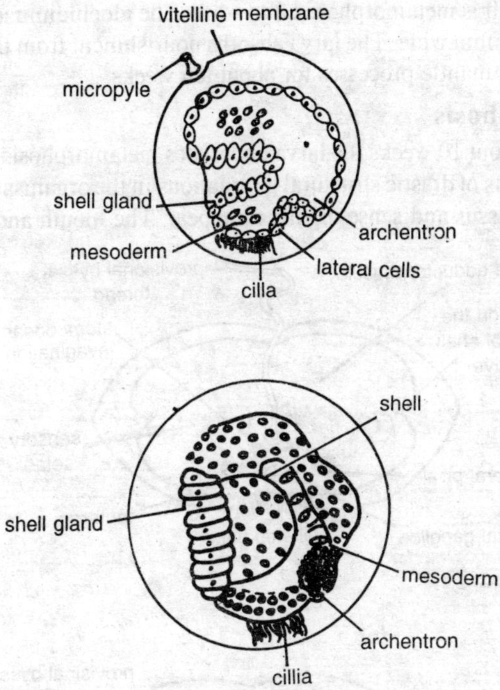

Fig. 6.21. *Unio.* An early and late embryo.

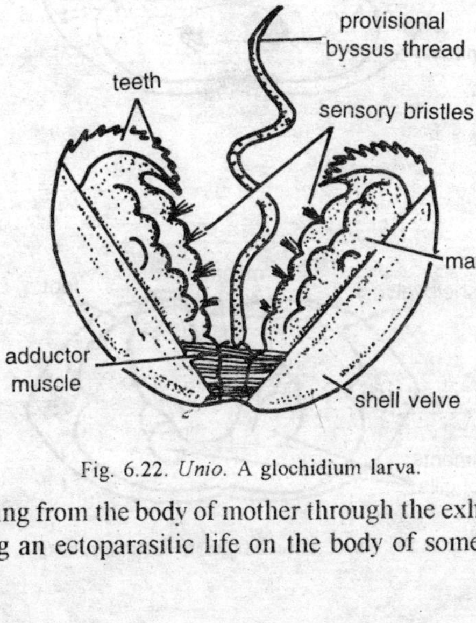

Fig. 6.22. *Unio.* A glochidium larva.

but on escaping from the body of mother through the exhalent siphon, it starts leading an ectoparasitic life on the body of some fish, which is persisted till it is metamorphosed into adult. The glochidium is a minute larva 0.1-0.5 mm wide. The larva absorbs nourishment from the body of host through mantle processes for about ten weeks.

Metamorphosis

After about 10 weeks the larva undergoes metamorphosis, which as usual, consists of drastic structural alternations in the organisation of the body. The byssus and sense organs disappear. The mouth and anus are

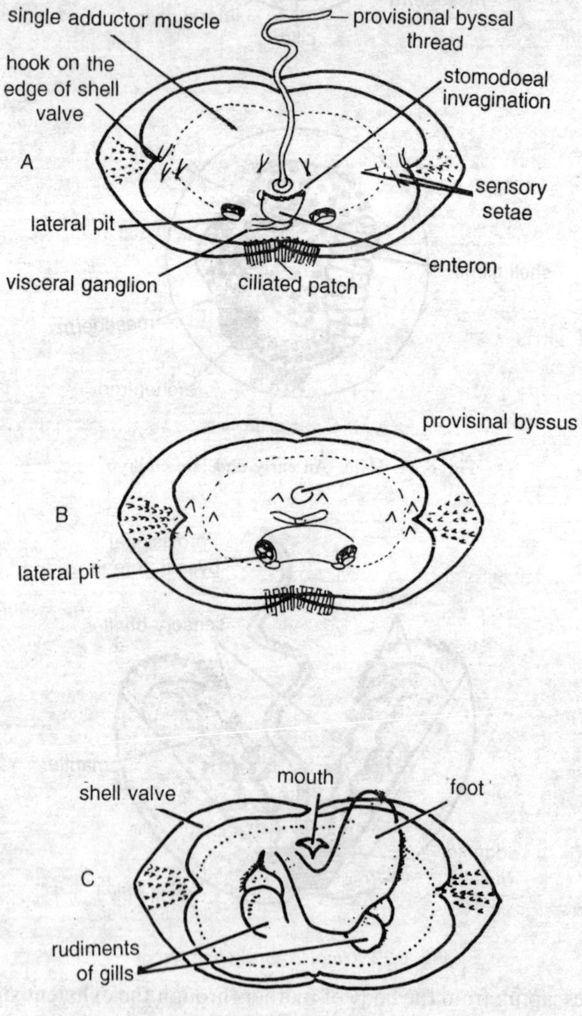

Fig. 6.23. *Unio*. Three stages in metamorphosis.

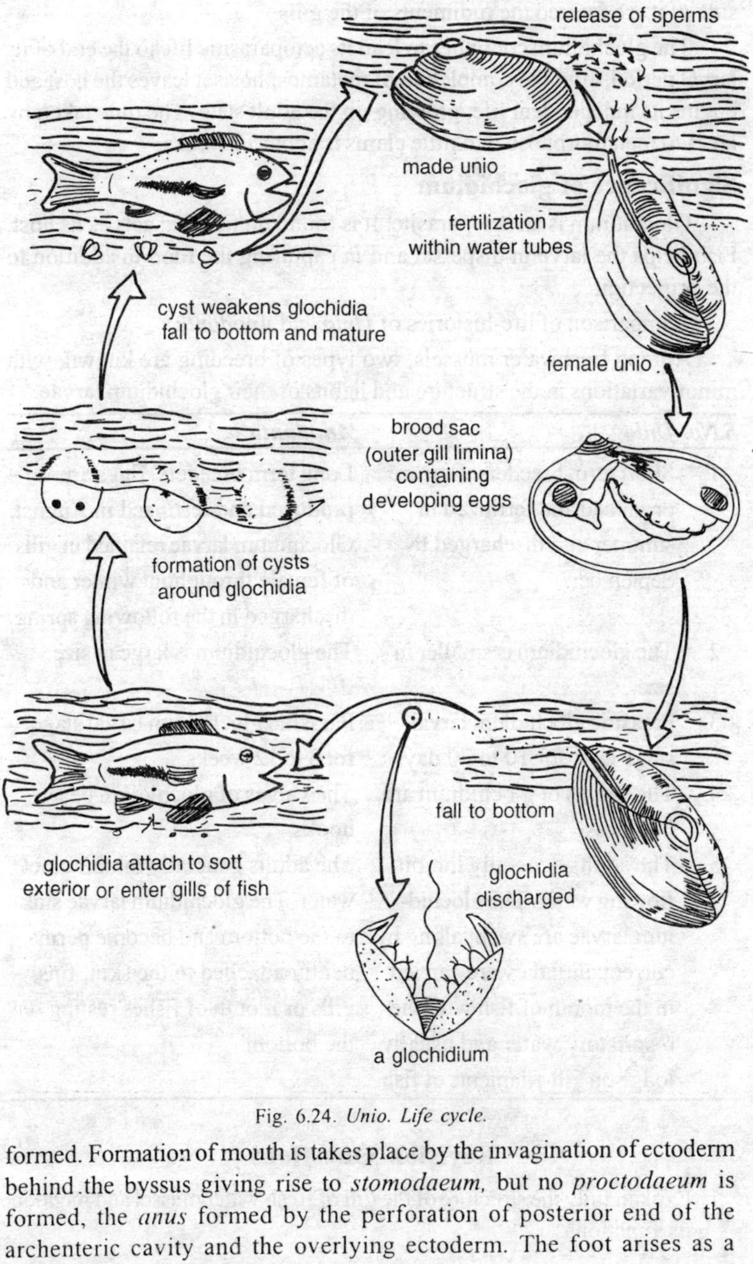

Fig. 6.24. *Unio. Life cycle.*

formed. Formation of mouth is takes place by the invagination of ectoderm behind the byssus giving rise to *stomodaeum,* but no *proctodaeum* is formed, the *anus* formed by the perforation of posterior end of the archenteric cavity and the overlying ectoderm. The foot arises as a projection from the mid-ventral surface behind the mouth and on either side of it are formed the rudiments of the gills.

The glochidium continues to lead its ectoparasitic life to the end of its larval period, after the completion of metamosphosis it leaves the host and begins its independent life, growing on the adult state. The time taken by larva to metamorphose into little clams is about 10 weeks.

Significance of glochidium

Glochidium is an ectoparasite. It is found on fish that acts as its host. Fish helps the larve in dispersal and in capturing the food in addition to the protection.

Comparison of life-histories of *Unio* and *Anodonta*.

Among freshwater mussels, two types of breeding are known, with minor variations in the structure and habits of their glochidium larvae.

S.No.	*Unio*	*Anodonta*
1.	Short term breeders. Egg are produced and fertilized in summer and discharged by September.	Long term breeders. Eggs are produced and fertilized in August. Glochidium larvae retained in gills of female throughout winter and discharged in the following spring.
2.	The glochidium is smaller in size.	The glochidium is large in size.
3.	Parasitic glochidium larval stage lasts for 10 to 70 days	Parasitic glochidium larval stage for 3 to 12 weeks.
4.	The valves of glochidium are hookless.	The valves of glochidium bear hooks.
5.	The adults generally inhibit flowing water. The glochidium larvae are swept along by current until they are carried in the mouth of fish with the respiratory water and usually lodge on gill-filaments of fish.	The adults generally inhibit quiet water. The glochidium larvae sink to the bottom and become permanently attached to the skin, fins, gills or mouth of fishes resting on the bottom.

Revision Questions

1. Explain fully the structure of the gill of fresh water mussel and mention its functions.

2. Describe the reproductive organs and development of *Lamellidens*.
3. Describe the digestive system and mode of feeding in *Unio*.
4. Draw a labelled diagram of circulatory system of *Unio*.
5. Describe the mechanism of respiration and the respiratory organs of *Unio*.
6. Describe the structure and metamorphosis of the glochidium larva.
7. Describe the reproductive organs of *Unio* and discuss the structure and significance of glochidium larva in its life-cycle.
8. Describe the respiratory organs of *Unio* and compare it with that of *Pila*.
9. Draw a neat labelled diagram of T.S. of *Unio* passing through gill region.
10. Give a comparative account of digestive system of *Pila* and *Unio*.
11. Draw neat and labelled diagrams of T.S. of *unio* passing through anterior, middle and posterior regions.
12. Write short notes on : (a) Mantle, (b) Keber's organ, (c) Glochidium larva.

7

Sepia

Sepia comes under Class Cephalopoda in which the foot is transformed into circum-oral appendages, completely surrounding the head, and the epipodium is modified as funnel or siphon or exhalent muscular tube.

SYSTEMATIC POSITION

Phylum	-	Mollusca
Class	-	Cephalopoda
Subclass	-	Dibranchiata
Order	-	Decapoda
Genus	-	*Sepia*

Distribution

Sepia is commonly called *cuttle fish*. It is cosmopolitan in its distribution, and is more common in warmer seas like Mediterranean, Indian ocean, Pacific etc.

Habits and Habitat

Sepia is a marine mollusc. It inhabits shallow waters usually living between stones and in crevices of rocks but often burrows in sand.

It is a good swimmer. The swimming movements are forwards and backwards by the fins and funnels. It is carnivorous and feeds on crustaceans and fishes. It is capable of colour changing. When followed by an enemy, it discharges ink into water from its ink gland. Sexes are separate. Breeding occurs in summer.

EXTERNAL FEATURES

Shape and size

Sepia has a fishy, bilaterally symmetrical and dorsoventrally flattened body. The anterior and posterior ends of the body, infact, represent the dorsal and ventral ends due to much elongation of dorso-ventral axis of the body. The body- colour is brownish and it is about 20 cm. in length.

Structure

The body is divisible into three parts:

1. Head
2. Neck
3. Trunk

1. Head

The head is almost globular in form. It bears a pair of large, highly developed eyes at the sides and mouth at its free end. The mouth is surrounded by five pairs of muscular processes. These are differentiated into four pairs of short and stout arms and one pair of long tentacles retractile into large pits at their bases. When extended, they are about 2/3 of the length of the body. They have strongly convex outer surface and flat inner surface. The inner surface bears four longitudinal rows of suckers.

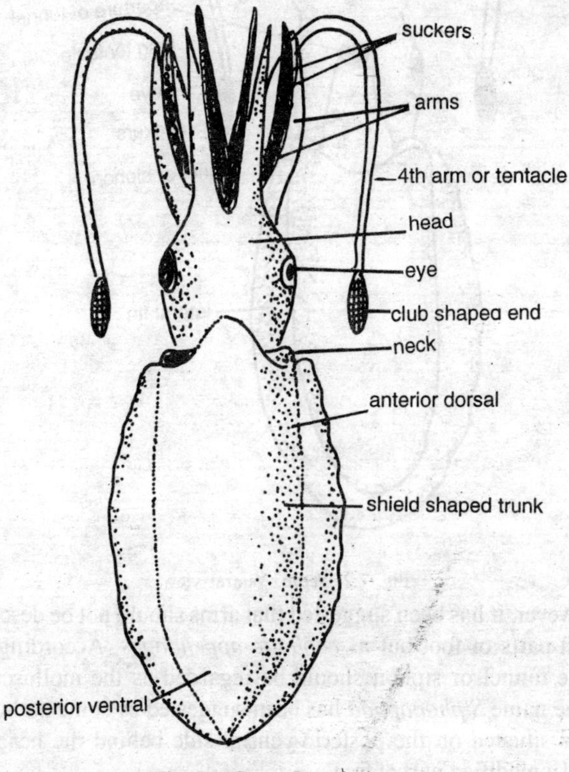

Fig. 7.1. *Sepia.* Dorsal view.

Each sucker is a muscular, shallow cup with a narrow, horny rim and supported on a short, thick stalk. The tentacles bear suckers on their expanded club-like tips only. The tentacles are used in the capture of prey and in copulation. The left arm of the fifth (posterior) pair is slightly modified in the male. The modification simply pertains to the loss of some of the suckers. This arm is called as *hectocotylized* and is used for transferring spermatophores into the mantle cavity of female. The arms, tentacles and the head proper are collectively called *cephalo-podium*. Some workers regard arms as the main part of the foot.

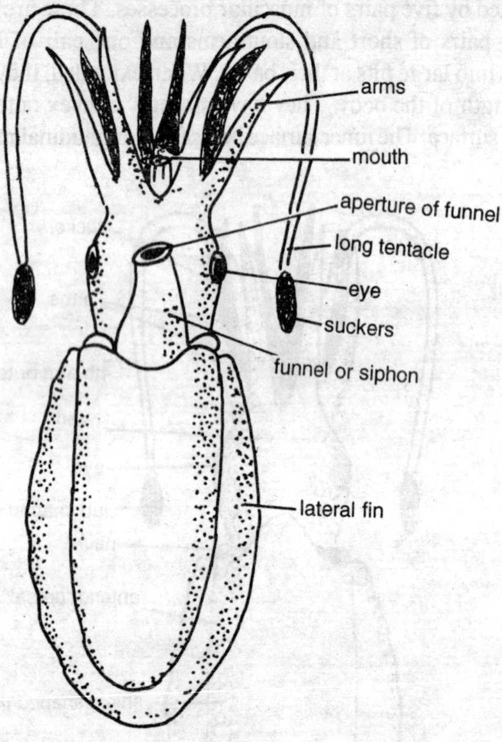

Fig. 7.2. *Sepia*. Ventral view.

However, it has been suggested that arms should not be described as modified parts of foot but as *cephalic appendages*. According to this view, the funnel or siphon should be regarded as the molluscan foot, hence the name *Siphonopoda* has been suggested by some authors. The funnel is situated on the posterio-ventral side behind the head and is regarded a modified part of foot.

2. Neck

It connects the head with trunk. It bears a funnel which opens to out side by a narrow aperture, and by a large aperture into mantle cavity.

3. Trunk

The trunk is elongated somewhat flattened and shield-shaped. Its broad base is directed orally and the pointed apex aboally. It is bordered by a narrow *fin* on either side which helps in locomotion at leisure. One side of trunk is darker, slightly convex and hard due to presence of shell inside.

Mantle

The trunk is covered by a thick fold of mantle,which terminates towards the oral surface in a ridge around the neck. The head can be partially retracted. Posteriorly the ridge forms a large cavity bounded by the mantle called the *mantle cavity* or *pallial cavity*, later contains the entire viscera.

Funnel

Below the head lies a large conical muscular tube, the funnel or siphon projecting beyond the neck. It internally communicates with the mantle cavity by a wide aperture, thus the funnel is the main out-let of the mantle cavity. A pair of cartilaginous knobs on the mantle fits into corresponding sockets on the posterior ventral surface of the funnel. The funnel not only carry to the exterior the foul water of respiration, the faecal matters and the products of excretory and reproductive organs, but also takes an important part in locomotion.

Skin

Flat contractile, pigment-containing cells, the *chromatophores* occur in the deeper layers of the integument over the entire surface. They have elastic membrane from which bundles of muscle fibres radiate to the surounding tissue. The chromatophores are of the three kinds reddish, yellowish -brown and orange. They can be detailed by the contraction of radiating muscles. The living *Sepia* can change its colour. Besides the chromatophores, a number of cells, the *iridocytes* are also present in the integument which cause the peculiar irridescence in addition to the vivid colouration caused by the chromatophores.

Shell

The shell is internal and is called the *cuttlebone* or *sepiostaire*. It is enclosed in a sac of mantle on the anterior surface of the trunk. It is secreted by the sac itself. It is flat and roughly oval in form with rounded oval end the *pro-ostracum* and a pointed aboral end, which has a sharp spine or *rostrum*.

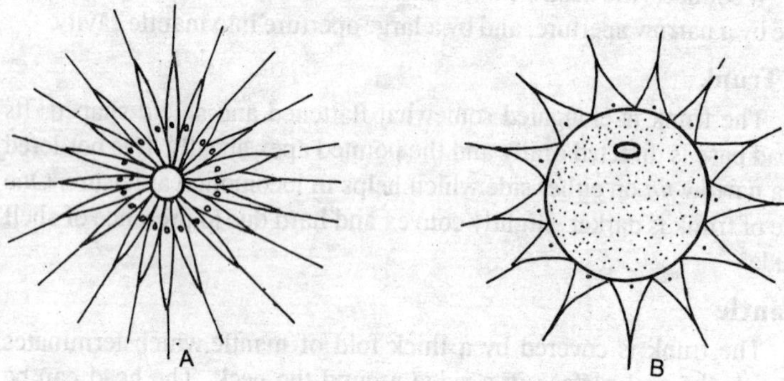

Fig. 7.3. *Sepia.* Chromatophore A—Contracted, B—Expanded.

The shell is non-living and consists of a large number of closely arranged calcareous laminae. The laminae enclose air-spaces between them to make the shell-light. Thus the shell besides acting as endoskeleton also serves as a float and help to maintain equilibrium of the body, being dorsal in position. The shells of dead *Sepia* are often found on the sea-coast thrown by the tides. These are often called the *sea-foam*.

Mantle cavity

It is a large space enclosed between the trunk and the mantle on the ventral side. It is closed on all the sides except the oral side, where it communicates with the exterior by two passages arround the neck, the *pallial aperture* and a large tube-like *funnel* or *siphon*.The funnel has a wide internal opening and a narrow external opening. Basally, on its ventral surface, the funnel bears a pair of cartilaginous sockets, called the *funnel* or *infundibular cartilages*. Opposite them, on the internal surface of the reflected mantle folds, are seen a pair of oval cartilaginous knobs, the *mantle cartilages*. Similarly, the dorsal surface of neck carries the *nuchal cartilage*, which fits against the dorsal cartilage on the mantle. By enlarging the mantle cavity, the *Sepia* draws water into it through the pallial aperture. When the mantle cavity is reduced, the mantle and infunfibular cartilages get interlocked to close the pallial aperture. Therefore, the water leaves the mantle cavity through the funnel.The funnel is guarded by a flap-like *valve* that opens outwards.

The whole of the mantal cavity is occupied by visceral mass having various organs. The digestive gland is situated anteriorly on either side of the retractor muscles of head and funnel. There is a pair of plume-shaped

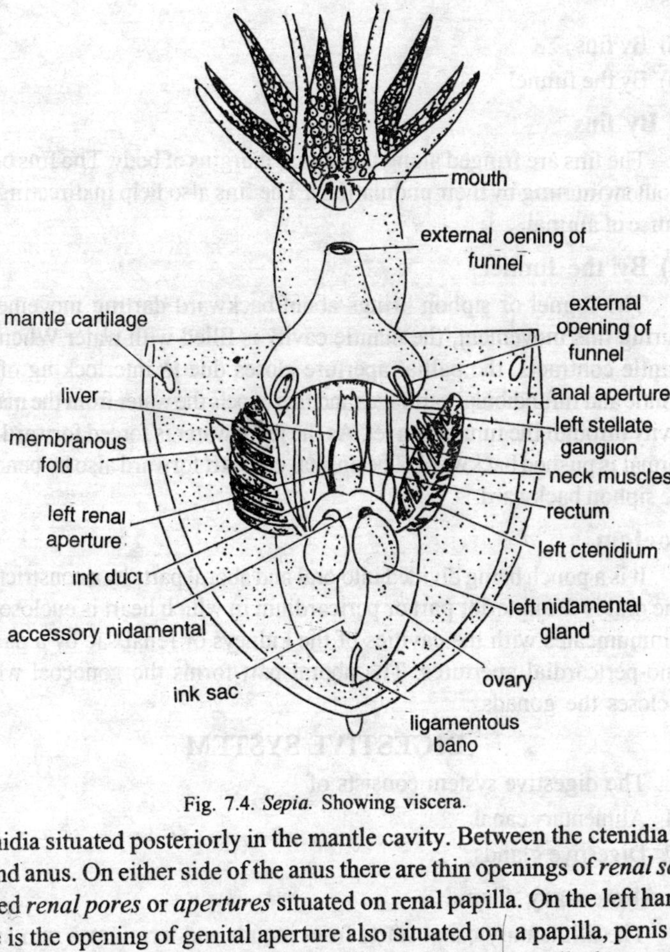

Fig. 7.4. *Sepia.* Showing viscera.

ctenidia situated posteriorly in the mantle cavity. Between the ctenidia is found anus. On either side of the anus there are thin openings of *renal sac* called *renal pores* or *apertures* situated on renal papilla. On the left hand side is the opening of genital aperture also situated on a papilla, penis is in the male and oviduct is found in female. On the mantle wall where neck joins the trunk are situated two large stellte ganglia one on each side. Posteriorly behind the *nidamental glands* is found an *ink sac* which contains ink made of melanin pigment. The duct of the ink-sac, runs forward ventral to the rectum and opens dorsally close to the anus. In male *Sepia*, testis is situated covered by an ink-sac while in female renal sac are invisible as they are hidden by nidamental glands and a pair of accessory nidamental glands.

Locomotion

Sepia exhibits two types of locomotion. They are:

(i) By fins
(ii) By the funnel

(i) By fins

The fins are fringed along the lateral margins of body. The fins bring about swimming by their undulations. The fins also help in directing the course of animal.

(ii) By the funnel

The funnel or siphon brings about backward darting movements. During this movement, the mantle cavity is filled with water. When the mantle contracts, the pallial aperture closes due to interlocking of the mantle and infundibular cartilages and this expels the water from the mantle cavity through the funnel as a jet. As the jet of water is forced forward, the animal is pushed backward. The animal can dart forward also by bending the siphon backward.

Coelom

It is a pouch being divided into oral and aboral parts by a constriction. The oral is the anterior part or pericardium in which heart is enclosed. It communicates with the cavities of the kidneys or renal sac by a pair of reno-pericardial apertures. The aboral part forms the gonocoel which encloses the gonads.

DIGESTIVE SYSTEM

The digestive system consists of

1. Alimentary canal
2. Digestive glands.

1. Alimentary Canal

The alimentary canal starts from mouth and terminates into anus. *Mouth* is present among the oral arms and is surrounded by a thin, fleshy

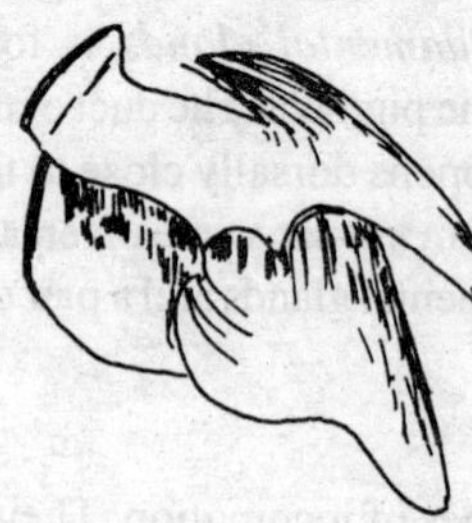

Fig. 7.5. *Sepia*. A Jaw.

circular lips beset with numerous papillae. Two *jaws* are located just within the lips. The jaws have somewhat the appearance of the beak of a parrot. The mouth leads into a large thick-walled, muscular *buccal cavity* or *pharynx*. Buccal cavity contains on its floor the *odontophore* and the *radula* bearing numerous minute, horny teeth. The buccal cavity leads into the *oesophagus*, which is a long narrow tube. It runs straight towards the aboral end, passing between two lobes of liver. The oesophagus opens into a rounded thick walled muscular bag, the *stomach*. The stomach is followed by the tubular intestine. At the junction of stomach and intestine, a wide curved sac opens into the alimentary canal. It is known as *caecum*.

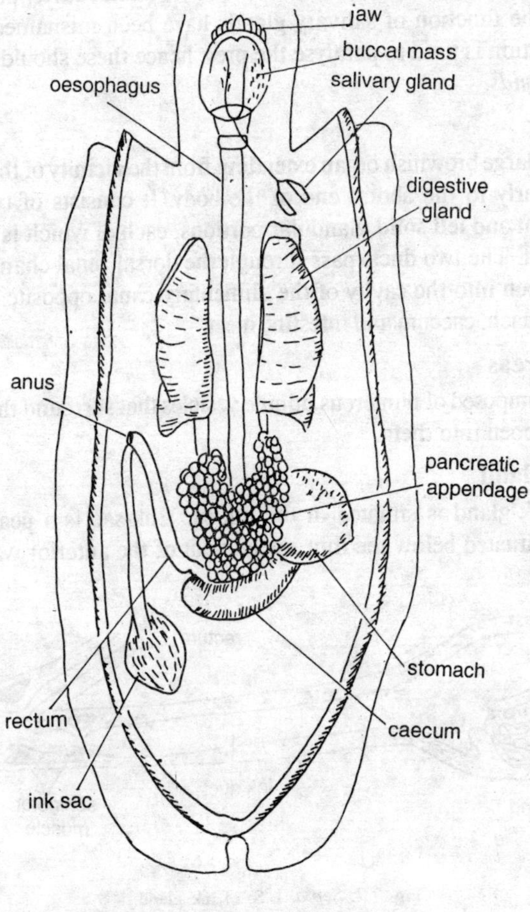

Fig. 7.6. *Sepia*. Alimentary Canal.

The intestine runs forward into a *rectum*. The rectum opens into the mantle cavity by *anus*. A pair of leaf-like *anal-valves* project at the sides of anus. Their function is uncertain.

2. Digestive glands

The glands associated with alimentary canal are :

(a) Salivary glands

A pair of anterior salivary glands is present within the buccal mass opening on either side of the radula. A pair of posterior salivary glands, lying in front of liver one on either side of oesophagus. A salivary duct leads inwards from each gland and joins its fellow from the opposite side to form median salivary duct which opens into buccal cavity at the tip of tongue. The function of salivary glands have been misnamed because their secretion is used to paralyse the prey, hence these should be called *poison glands*.

(b) Liver

It is a large brownish organ extending from the vicinity of the salivary glands nearly to the aboral end of the body. It consists of two partly united right and left solid glandular portions, each of which is provided with a duct. The two ducts pass through the dorsal renal chambers and unite to open into the cavity of the alimentary canal opposite the point where stomach, caecum and intestine meet.

(c) Pancreas

It is composed of numerous minute vesicles that surround the hepatic ducts and open into them.

(d) Ink gland

An ink-gland is situated in the ink-sac. Ink-sac is a pear-shaped structure situated below the thin integument of the anterior wall of the

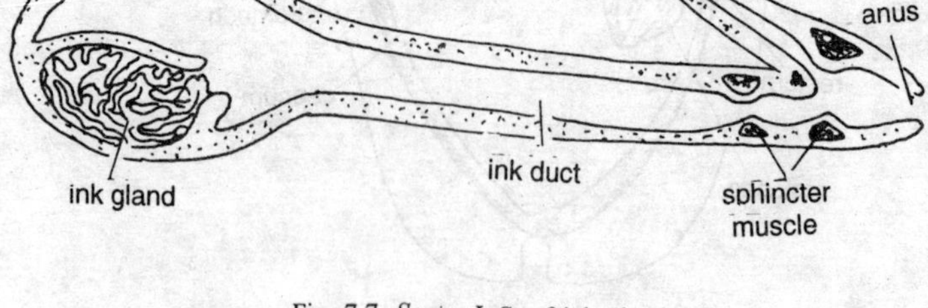

Fig. 7.7. *Sepia*. L.S. of ink gland.

mantle cavity. It opens by a duct dorsally into the rectum close to the anus. The ink-gland secretes a brown or black fluid or ink. It contains a high concentration of melanin pigment and is stored in the ink sac. When startled, the cuttle fish discharges the ink, which mixes with water in the mantle cavity and passes out through the funnel as a dark-cloud, under the cover of which the animal escapes from an enemy or approaches a prey.

Food

Food of *Sepia* consists of crustaceans e.g., prawns and shrimps; molluses and small fishes. Prey is caught by the protrusion of the tentacles. The prey is attached to the tentacles by the suckers and drawn in within the reach of the arm. Jaws which are situated in the mouth help in crushing the food, radula in grasping the flesh and in swallowing. A pair of salivary glands is also situated in *Sepia*. These glands produce a poison which kills the prey.

Digestion

The food caught by tentacles is paralysed by salivary glands, crushed into pieces by jaws swallowed with the help of radula, reaches the stomach where it is mixed with the digestive fluids from the liver and pancreas. The digestion is completed in the caecum where liquid products are absorbed while undigested food is passed into the intestine where some absorption may take place. The undigested food material thrown out of the anus. The hepato-pancreas secretes digestive enzymes and serves for the absorption and storage of the food and removal of waste products. The food stored in the liver is not directly absorbed but received through the blood.

RESPIRATORY SYSTEM

The respiratory organs are paired, large, plume-shaped *ctenidia* or *gills*, lying in the mantle cavity, one on either lateral sides. Each ctenidium is bipinnate, with numerous delicate lamellae on either side of a central axis. The respiratory surface of the lamellae is increased due to the presence of various folds on the lamellae. Internally the lamellae are not completely in contact with each other. Each gill receives venous blood through an afferent branchial vessel from the branchial heart of its side. The blood passes in a system of minute branches through the lamellae and is finally collected into an efferent branchial vessel leading to the auricle.

Respiratory mechanism

The muscular mantle rhythmically contracts and expands, thus the mantle cavity alternately increases and decreases. When the mantle cavity enlarges, fresh water flows into it through the pallial aperture. This water passes over the ctenidia. Exchange of gases occur in the vascular lamellae. Carbon dioxide is diffused out in the water and when the mantle contracts, the water is expelled out through the funnel.

CIRCULATORY SYSTEM

The circulatory system is well developed with a complete separation of venous and arterial blood. It consists of heart, blood vessels and a system of capillaries and blood.

Heart

There are three hearts; a large *systemic* or *arterial heart* and a pair of small *branchial* or *venous hearts*.

The systemic or arterial heart lies in the middle of the visceral mass enclosed in the pericardium. It is three chambered, having a large, median, thick-walled ventricle and two small lateral, thin-walled auricles. The ventricle is divided into two lobes by a constriction. It supplies arterial blood through a large *oral* or *cephalic aorta* and a small *aboral* or *posterior aorta* to the anterior and posterior regions of body. The auricles receive oxygenated

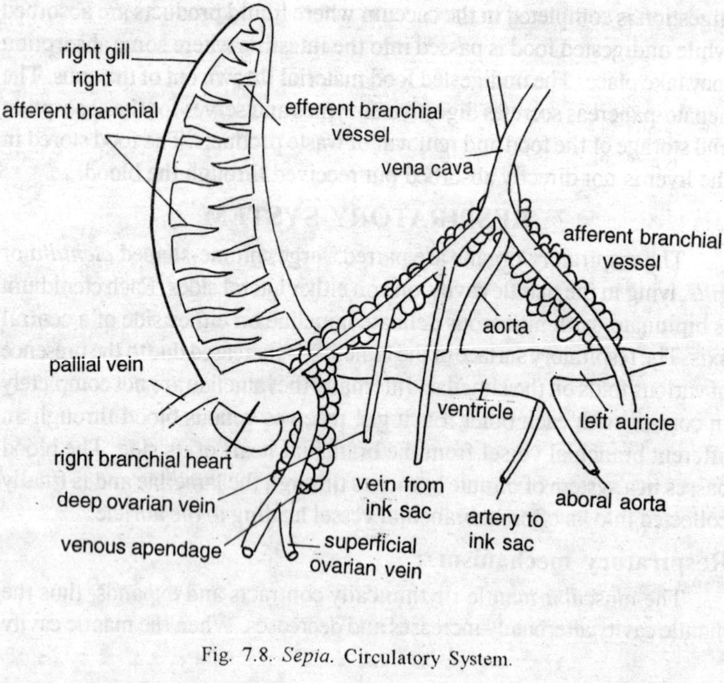

Fig. 7.8. *Sepia*. Circulatory System.

blood by the *efferent branchial veins*. This blood passes into the ventricle and is send out via aortae.

The branchial hearts lie one at the base of each ctenidium. They are simple contractile sacs that receive and supply deoxygenated blood by afferent branchial veins. A rounded glandular body called *appendage* is attached to each branchial heart.

Blood vessels

The oral and aboral aortae divides into small branches, the *arteries*, which divide and redivide to form very fine, microscopic vessels, the *capillaries*. The capillaries reunite to form large vessels, the *veins*. A large median *vena cava* returns blood from the head region. In front of the rectum, it bifurcates into right and left afferent branchial veins, which run through the corresponding renal organs to the base of ctenidia. Here they dilate to form branchial hearts.

A *pallial vein* fom the mantle and an *abdominal vein* from the posterior region pour the venous blood into each branchial heart. The righ vena-cava or right branchial vein also receives a vein from ink-gland and gonads. Each branchial heart pumps the blood to the ctenidium of its side by an afferent branchial vein, which runs through the axis of gills and gives off branches. The oxygenated blood from the ctenidium is returned by an efferent branchial vein first to the auricle and then to the ventricle.

Blood

Blood consists of the plasma containing *haemocyanin*, dissolved in it. Colourless *amoebocytes* floating in it. It is colourless when deoxygenated and light blue when oxygenated.

EXCRETORY SYSTEM

The process of excretion is affected by a *kidney* or *renal sac* which consists of three thin-walled chambers communicating with one another. Out of them one is dorsal and median while the other two are ventral and symmetrically situated on either side of the rectum. The two ventral parts of the kidney open to the outside by the *renal apertures* situated on the renal papillae. Through each ventral chamber passes the corresponding branchial vein of vena-cava. The vein is attached to the dorsal walls of the ventral renal chambers and are covered by glandular epithelium which is excretory in nature. The nitrogenous waste products are extracted from blood by glandular walls and reach into the renal sac. The dorsal chamber encloses the pancreatic follicles covering and opening into the ducts of the digestive gland. They are richly vascular and are said to serve an excretory function. The nitrogenous excretory matter, in the cavities of renal sac, is in the form of guanin which is poured into the mantle cavity.

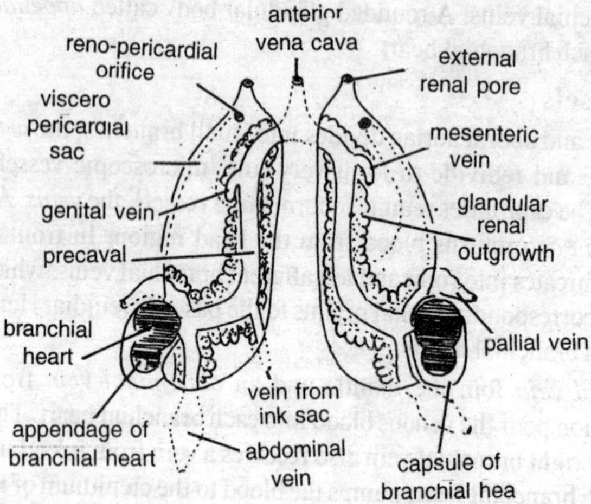

Fig. 7.9. *Sepia*. Excretory System.

NERVOUS SYSTEM

The nervous system shows a higher grade of organisation. The nervous system consists of ganglia, connectives and commissures. Following are the ganglia :

1. Cerebral ganglia
2. Pedal ganglia
3. Pleuro-visceral ganglia
4. Stellate ganglion or pallial ganglion
5. Gastric ganglion.

1. Cerebral ganglia

Paired cerebral ganglia are found together into a rounded mass lying in a hollow of the cranial cartilage and covered over anteriorly by a strong fibrous membrane. The brain gives off from the sides a pair of very short, thick *optic nerves* that at once expand into large *optic ganglia* lying in close contact with eyes. A small *olfactory ganglion* lies close to the optic ganglion and sends a fine olfactory nerve to the olfactory pits. The cerebral ganglia are connected with the *superior-buccal ganglia* anteriorly by thin *cerebro-buccal commissures*. The superior buccal ganglia are connected with a pair of closely united *inferior buccal ganglia* by a pair of *circumoesophageal connectives* join the brain ganglion with pedal and pleuro-visecral ganglia that lie posterior to oesophagus. A pair of nerves extend from the brain to the statocysts by way of the pedal ganglia.

2. Pedal ganglia

The pedal ganglia are fused into a mass. They send 10 branchial nerves to the arms. Each nerve expands into a *branchial ganglion* at the base of the arm it supplies. The pedal ganglia also give off a pair of *infundibular nerves* to the funnel. The pedal ganglia are connected with the superior buccal ganglia by *bucco-pedal connectives* and with the cerebral ganglia by *cerebro-pedal connectives*.

3. Pleuro-visceral ganglia

These are also united into a single mass being in immediate contact with the pedal ganglia behind the oesophagus. The pleurovisceral ganglia

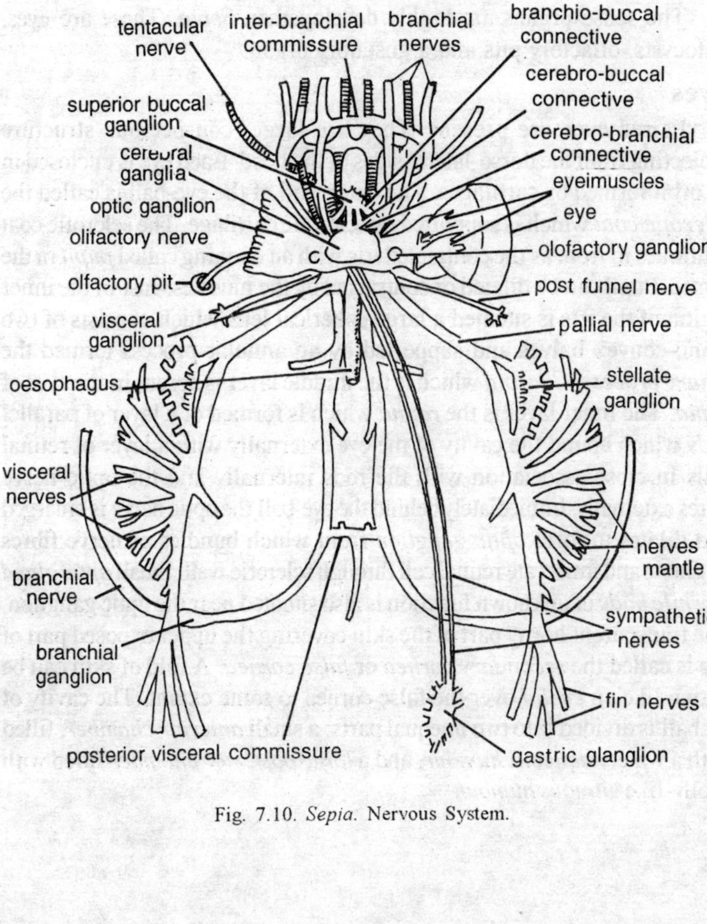

Fig. 7.10. *Sepia.* Nervous System.

give off two *visceral nerves* innervating the various internal organs and forming a *visceral loop* from which arise a pair of *branchial nerves*, having each a *branchial ganglion* at the base of the ctenidium. The pleuro-visceral ganglia also give off two very stout *pallial nerves* which innervate the mantle.

4. Pallial ganglion

Each pallial nerve runs through the neck to the inner surface of the mantle cavity and expand into a large, flat, pallial or stellate ganglion which is visible in front of the ctenidium when the mantle cavity is exposed.

5. Gastric ganglion

A pair of sympathetic nerves arise from the inferior buccal ganglia and extend aborally along the oesophagus to the stomach, where they join to form the gastric ganglion.

SENSE ORGANS

The sense organs are highly developed in *Sepia*. These are eyes, statocysts, olfactory pits and a gustatory organ.

Eyes

Paired eyes are present. They are large, conspecious structure projecting from the dorso-lateral sides of the head. Each eye is enclosed in an orbit formed by cartilages. The outer wall of the eye-ball is called the *sclerotic coat* which is supported by sclerotic cartilage. The sclerotic coat continues in front as the contracted iris with an opening called *pupil* in the centre. Pupil can be dilated or contracted by the muscles. Just in the inner region of the iris is situated a large spherical lens which consists of two plano-convex halves and supported by an annular process termed the *ciliary process*. *Choroid* which is the middle layer is absent in the eyes of *Sepia*. The inner layer is the *retina* which is formed of a layer of parallel *rods* which bound the cavity of the eye externally with a layer of retinal cells in close association with the rods internally and the optic nerve fibres externally. Immediately behind the eye-ball the optic nerve is enlarged and dilates to form *optic ganglion* from which bundles of nerve fibres originate and innervate retinal cell through sclerotic wall. Small *optic gland* or *white body* of unknown function is also situated near the optic ganglion. The transparent horny part of the skin covering the upper exposed part of eye is called the *secondary cornea* or *false cornea*. A fold of skin can be drawn like an *eyelid* over the false cornea to some extant. The cavity of eyeball is divided into two unequal parts: a small *anterior chamber*, filled with a watery *aqueous humour*, and a *large posterior chamber* filled with a jelly-like *vitreous humour*.

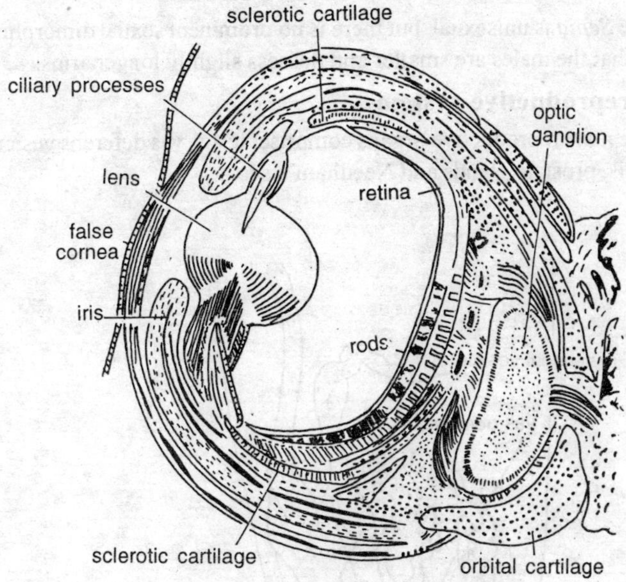

Fig. 7.11. *Sepia.* V.S. of eye.

Thus the eye of *Sepia* resembles those of vertebrate eye. However, the vertebrate eye is formed by the outgrowth of brain whereas the cephelopod eye is formed by an ectodermal invagination.

Statocysts

A pair of statocysts, which are organs of equilibrium, lie ventral to the pleuro-visceral ganglia. Each statocyst remains enclosed in cranial cartilage. A statocyst is a rounded sac, about 3 mm. in diameter. The inner surface is uneven and covered by flat cells, except on posterior side where two spots occur. These are *crists*, a *statica* and *macula statica*.These have large processes at the free ends and the other end produced into fine fibril that forms the *statocyst nerve*. A large dense *statolith* lies within the statocyst attached to the macula.

Olfactory pits

Small ciliated olfactory pits are situated posterior to the eyes. The sensory cells of the pits are innervated from small olfactory ganglia lying close to the optic ganglia.

Gustatory organs

A small elevation, covered with papillae on the floor of buccal cavity just infront of the odontophore, is said to be the organ of taste.

REPRODUCTIVE SYSTEM

The *Sepia* is unisexual, but there is no prominent sexual dimorphism except that the males are smaller and possess slightly longer arms.

Male reproductive system

The male reproductive organs comprise: testis, vas deferens vesicula seminalis, prostate glands and Needham's sac.

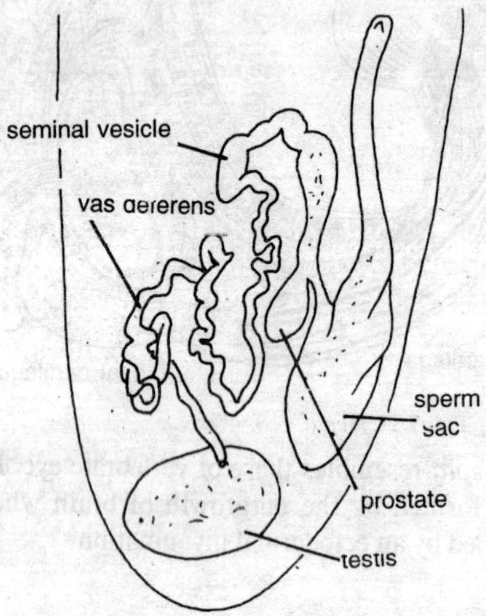

Fig. 7.12. *Sepia.* Male reproductive organs.

(i) Testis

The testis is large, oval, yellowish and saccular structure lies near the apex of visceral mass. It is enclosed in a coelomic sac and opens into it by a small aperture on its ventral side. The sperms produced in the testis are passed into the coelomic sac. The coelomic sac leads on the left side into a vas deferens.

(ii) Vas deferens

It is a narrow coiled tube, it leads into seminal vesicle.

(iii) Vesicula seminalis

It is a long tube. Here the sperms are rolled up into long and narrow bundles enclosed in elaborate chitinous capsules, called the *spermatophore*. A spermatophore is like an automatic bomb, at one end it has a complex spring-like arrangement which ruptures its wall as soon as transferred to water.

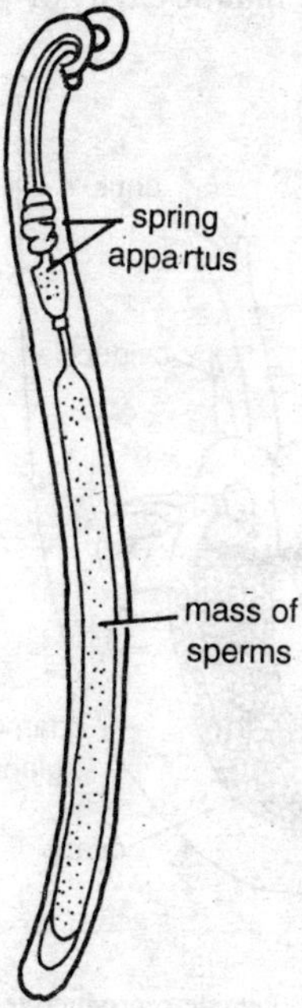

Fig. 7.13. *Sepia*. A spermatophore.

(iv) Prostate gland

The terminal end of vesicula seminalis gives off two blind folds, one of them is called the prostate gland and other is caecum.

(v) Needham's sac

The seminal vesicle terminates into a wide reservior, the Needham's sac lying on a papilla or the penis to the left of the anus.

Female reproductive system

The female reproductive organs comprise: ovary, oviduct, Nidamental glands and acessory Nidamental glands.

(i) Ovary

A large ovoid ovary situated, like the testis in the male in aboral region of the trunk. It is enclosed in a similar coelomic sac. It has an axial swelling that bears numerous follicles. Each follicle with a singel ovum supported on a stalk.

(ii) Oviduct

A wide thin-walled oviduct arises from the capsule or coelomic sac and runs orally to open into the mantle cavity by genital aperture.

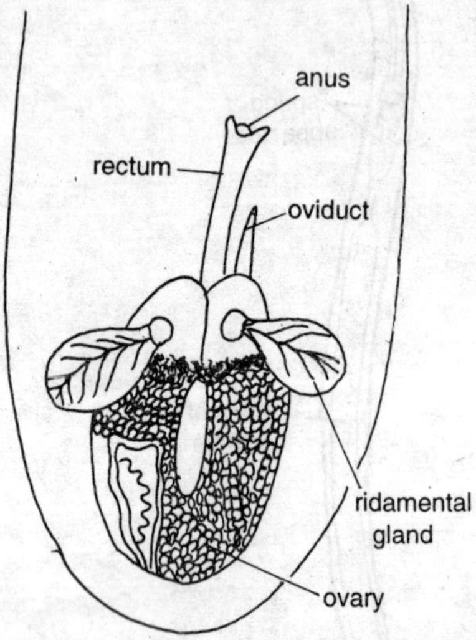

Fig. 7.14. *Sepia.* Female reproductive organs.

(iii) Nidamental glands

A pair of large, flat oval bodies lie on the anterior wall of the mantle cavity on the sides of ink-gland. These glands secrete a viscid material by means of which the eggs when deposits, adhere together is masses.

(iv) Accessory nidamental glands

A glandular mass of orange colour known as accessory nidamental glands lie at the sides and around the anterior sides of the nidamental glands. They open into the mantle cavity by a number of minute pores. Their function is not clear.

Copulation, fertilization and egg laying

In *Sepia* the 5th left arm of the male is *hectocotylized* i.e. modified for copulation by reduction of some of the basal rows of suckers. The male inserts it in its own mantle cavity to extract the spermatophores. Later on, this arm is inserted into the mantle cavity of the female and the spermatophores are deposited in *bursa copulatrix*, which is a modified part of the funnel. The spermatophores are burst by the spring apparatus

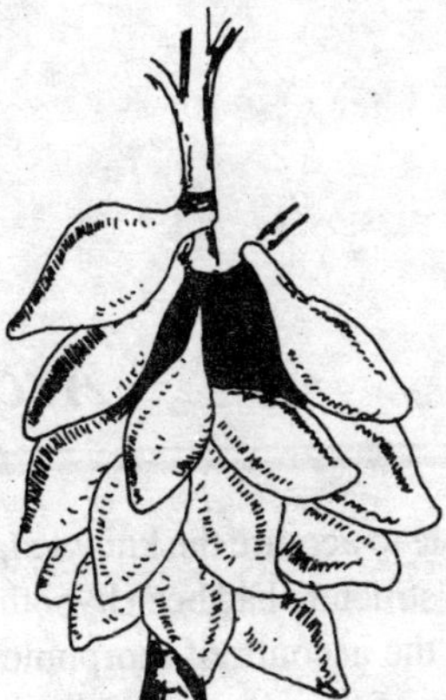

Fig. 7.15. *Sepia*. An egg mass.

in them, and release the sperms after copulation. When ripe ova are passed out side, one by one from the ovary through the coelomic funnel, they fertilized. The viscid secretion of nidamental gland covers the eggs and cement them into a mass. These masses of eggs are attached by stalks to the twig of marine plants or weed or some other object

The development is direct without larval stage.

Revision Questions

1. Describe the nervous system of *Sepia*.
2. Describe the digestive system and feeding mechanism of *Sepia*.
3. Draw neat and labelled diagram of mantle cavity showing viscera in *Sepia*.
4. Draw a labelled diagram of V.S. of eye of *Sepia*.

8

Archimollusca

It appears almost impossible to acquire the knowledge about the primitive form of mollusc, yet such a structure had been hypothetically reconstructed by careful observation on the account of morphological and anatomical researches. The description of the ancestor mollusc is based on the work of several workers. But largely it represents the views of *Yonge*. No living mollusc, not even *Neopilina*, is like this ideal hypothetical ancestor.

Habits and habitat

The ancestral mollusc was a inhabitant of Pre-Cambrian oceans where it moved about over rocks and other types of hard bottom in shallow water.

EXTERNAL FEATURES

The archimollusc had a bilaterally symmetrical body which was dorsally covered by a low conical shell. It measured 2.5 cm. in length. Mouth was anteriorly, and anus posteriorly placed.

Head

It had a well defined head, bearing one pair of tentacles, with one pair of eyes or ocelli at their bases.The mouth occupied the ventral position.

Foot

It had a ventral creeping foot with powerful and well-developed muscular sole. Around the base of foot runs an epipodial fringe,beset with delicate tactile tentacles.

Visceral mass

The dorsal and posterior part of the body raised like a hump. The hump formed the visceral mass which included the vital organs of the body.

Mantle

A fold of soft integument, the *mantle* or *pallium* hang down around the body from dorsal side enclosing mantle cavity and communicating out side between free edges of foot.

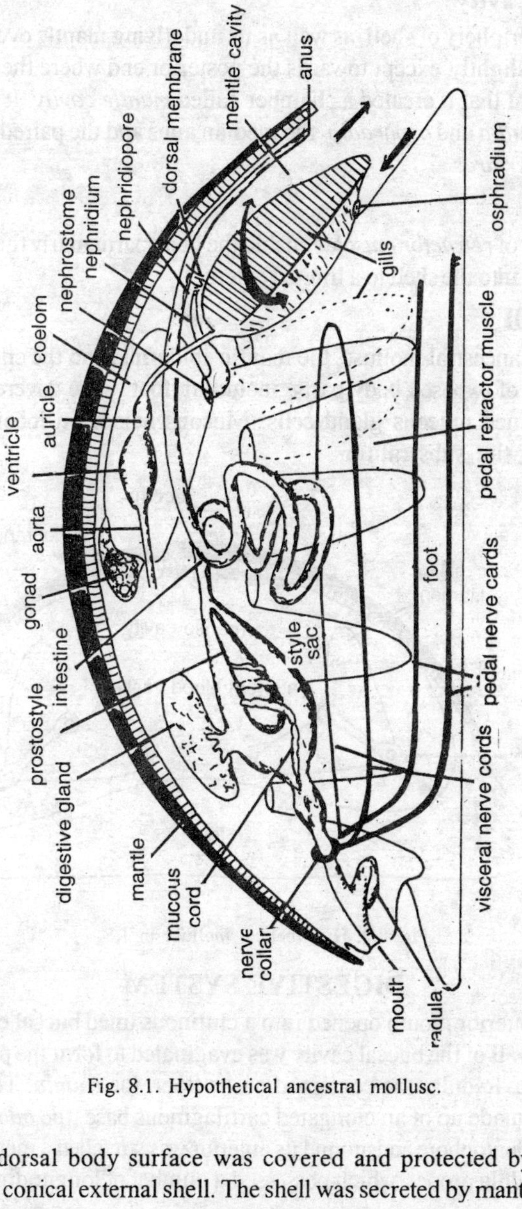

Fig. 8.1. Hypothetical ancestral mollusc.

Shell

The dorsal body surface was covered and protected by an oval, convex or conical external shell. The shell was secreted by mantle made of chitinous matrix and calcium carbonate. The shell provided support to the mantle cavity and attachment to dorso-ventral muscles.

Mantle Cavity

The periphery of shell, as well as its underlying mantle overhung the body only slightly, except towards the posterior end where the overhung was so great that it created a chamber called *mantle cavity*. It contained paired *ctenidia* and *osphradia,* the median anus and the paired *renal* and *genital apertures.*

Muscles

A pair of *retractor muscles* enable the body particularly the foot to be withdrawn into the shell to a limited extent.

Body Wall

In the ancestral mollusc, the mantle epidermis and the epidermis of remainder of exposed body parts, including foot, were covered by cilia and contained mucous gland cells. Mucous helped in locomotion as lubricated the substratum.

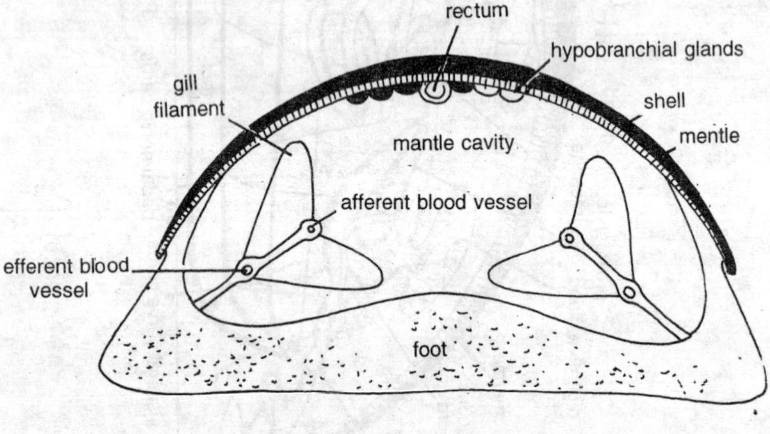

Fig. 8.2. Hypothetical mollusc in T.S.

DIGESTIVE SYSTEM

The anterior mouth opened into a chitinous lined buccal cavity. The posterior wall of the buccal cavity was evaginated to form the pocket-like radular sac. Redular sac contained on its floor, the *radula*. The radular apparatus made up of an elongated cartilaginous base, the *odontophore*. Over the odontophore and around its anterior was stretched a membraneous belt, the radula proper, which possessed a number of longitudinal rows of chitinous teeth. *Protractor* and *retractor muscles* were attached to the odontophore. The odontophore could be thrust out of mouth.

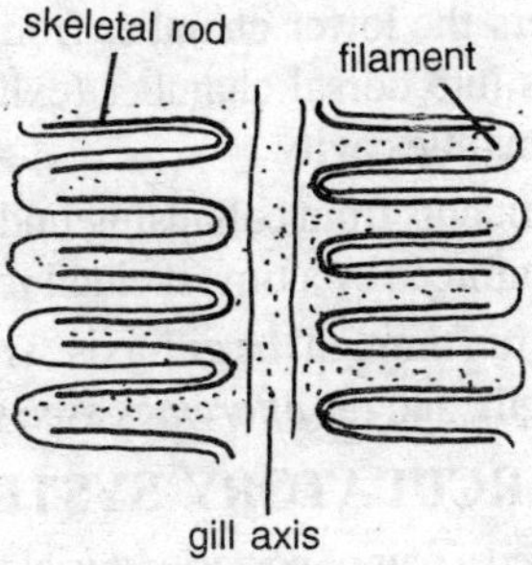

Fig. 8.3. Frontal Section through primitive gill.

A pair of *salivary glands* opened into the anterior dorsal wall of the buccal cavity. These glands secrete mucous. Buccal cavity followed by oesophagus which posteriorly opened into *stomach*. The stomach of archimollusc was complicated. Its anterior region was broad and hemispherical where as posterior end tapered, called *style sac*, led into the *intestine*. The anterior region of stomach was lined with chitin except for a ciliated, ridged sorting region and the entrance point for two ducts from a pair of lateral *digestive glands*.

The long coiled intestine functioned largely in the formation of faecal pellets. The *anus* opened mid-ventrally at the posterior margin of the mantle cavity and the wastes were swept away by the exhalent current.

In all probability, the ancestral mollusc was a microphagous herbivore that fed on fine algae growing on rocks in shallow water.

Coelom

The coelomic cavity was located in the mid-dorsal region of the body. The cavity surrounded the heart dorsally and a portion of intestine ventrally. Thus the coelom represented both-a pericardial and a perivisceral coelom.

RESPIRATORY SYSTEM

The gills of ancestral mollusc were probably like those of primitive living gastropods and bivalves. Such a gill consists of a long flattened axis projecting from anterior wall of the mantle cavity and contains blood vessels, muscles and nerves. To each side of the broad surfce of the axis are attached flattened triangular or wedge-shaped filaments, which alternate in position with those filaments on opposite side of the axis.

The gills are located on opposite sides of the mantle cavity and are held in position by a ventral and a dorsal membranes. Functionally, the position of the gills divides the mantle cavity into an upper and lower chambers. Water enters the lower chamber from posterior side, passes upwards through gills into dorsal chamber (exhalent) and then moves posteriorly out of the mantle cavity.

To prevent the filaments from collapsing under the flow of water,the frontal margin was reinforced by a tiny, skeletal rod made up of chitin.

Two blood vessels run through the gill-axis. The *afferent vessel*, which carried blood into the gill, and the *efferent vessel*, which draind the gill.

CIRCULATORY SYSTEM

The heart comprised a pair of posterior auricles and a single muscular anterior ventricle. The auricles received oxygenated blood from the gills and then passed it into the ventricle. Ventricle pumped blood through a single aorta. the aorta branched and rebranched into smaller blood vessels that delivered blood into sinuses, where the tissues bathed directly. Thus the circulatory system was of open type. From sinuses blood trawelled to the gills and back to the auricles. Plasma contained colourless *amoebocytes* as well as the respiratory pigment called *haemocyanin*.

EXCRETORY SYSTEM

The excretory organs of the archimollusc probably consisted of a pair of tubular metanephridia, commonly called *kidneys* in living species. Each nephridium opened at one end into the pericardium through a *nephrostome* that was located laterally to the intestine.The pericardial coelom received wastes from two sources. The heart wall deliver a filterate from blood, and glands in the pericardium secreted waste into the coelom. The pericardial fluid then passed through the nephrostome into the nephridium. The nephridium discharged into the mantle cavity through the *nephridiopore*.

NERVOUS SYSTEM

Nervous system was composed of a *nerve ring* around the oesophagus which may or may not had ganglia. From this nerve ring arose two pairs of nerve trunks running longitudinally through the body with ganglionic cells distributed along the entire length. The ventral pair represented the *pedal cords* and dorsal pair to the *visceral cords* innervating the mantle and visceral organs.

SENSE ORGANS

Sense organs consisted of paired *tentacles*, *eyespots*, a pair of *statocysts* in front and two *ospharadia*. The ospharadia were the patches of sensory epithelium, located at the bases of gills to test the quality of water. An *epipodium* running around the base of foot and bearing tactile and probably *chaemoreceptive tentacles.*

REPRODUCTIVE SYSTEM

The sexes were united. The gonads were located in the coelom and gametes were discharged through the kidneys, entering by way of the pericardial cavity.

Development

Fertilization was external in sea water. The eggs were alecithal and without protective capsules. The cleavage was spiral. Gastrulation by epiboly or invagination or both. The gastrula developed into a trochophore larva. Without passing through veliger larva, trochophore developed into adult.

9

Foot in Mollusca

In Mollusca, the ventral side of the body, is characterised by the pronounced development of its musculature, which enables the animal to creep. This structure is known as *foot*. A fleshy foot, provided with a flat sole, is suited for such a locomotion. The foot is quite distinct from rest of the body, especially from the head.

The flat form of the foot with a sole for creeping must be considered the primitive form, such a foot may be found in *Chitons* among the class Amphineura, but it is variously modified among various classes of Phylum Mollusca to perform diverse types of functions i.e., swimming, attachment, digging and so on. The foot and its appendages get their nerve-supply from the pedal ganglia or the pedal nerve cord.

The foot may be distinguished into various parts as follows:

(i) *Anterior Propodium* : It is anterior part which is distinctly marked from rest of foot.

(ii) *Metapodium* : It is the posterior part of the foot and carries operculum.

(iii) *Mesopodium* : It is the central portion between the first two.

Apart from this, sometimes there occurs expanded lobe-like margins of the ventral side of foot called as *parapodium,* which can be folded and united over the head. Sometimes there are paired, prominant ridges arising from the sides of the base of foot and called *epipodium*, which are provided with sensory tentacles and eye-sports. In cephalopods the foot becomes modified into the *tentacles, suckess* and *siphons.*

Variations of foot in different groups of Mollusca is as follow :

1. Foot in Monoplacophora

In *Neopilina* the ventral surface bears a prominant foot in the centre. The foot shows a primitive condition. It is broad, circular, with a flat sole on which the animal creeps by muscular waves.

2. Foot in Amphineura

Class Amphineura is divided in two subclasses. Polyplacophora and Aplacophora.

(i) *Polyplacophora.* It possesses most primitive type of foot, which is large muscular organ with a flat. creeping surface extending along the whole length of body on its ventral surface. The foot is not differentiated into different parts. The example is *Chiton*. Besides performing locomotory function of creeping, it also serves as an organ of adhesion to the surface of rocks, when at rest.

(ii) *Aplacophora.* The foot is vestigial and is represented by a ciliary ridge running along the base of narrow, medium, ventral groove, extending between the mouth and cloacal aperture. In some forms like *Chaetoderma* both, the foot and ventral groove are absent.

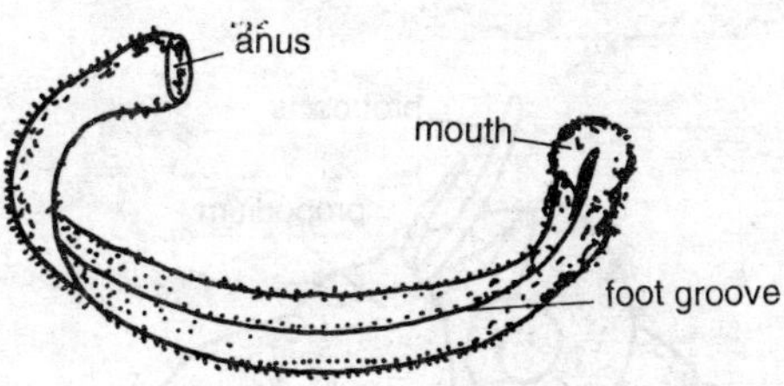

Fig. 9.1. Foot in *Chaetoderma*.

3. Foot in Scaphopoda

Foot is almost cylindrical or conical as in *Dentalium* and projects downward into the mantle cavity through the lower aperture of which it can be protruded out to serve as the digging organ. Engorged with blood, it serves as an anchor. The free end of the foot is conical and trilobed carrying on each lateral side a wing-like fold or pleat which is comparable to the epipodium. A groove runs along the middle line of foot.

In *Siphonodentalium* both, the groove and lateral folds are absent and the anterior end of foot bears a contractile disc with papillated margins, while in *Pulsellum,* a filiform tenacle springs from the middle of the disc.

4. Foot in Gastropoda

The foot of gastropods contains scattered unicellular as well as aggregated multicellular glands which secrete mucous. These glands are called *pedal glands*. The pedal glands include:

(i) *Labial gland* in a groove at the anterior end,

(ii) *Supra pedal gland* between head and anterior end of foot.

(iii) *Sole gland* opening mid-ventrally in the middle of foot and,

(iv) *Posterior glands* at the hind end of foot.

A. Foot in subclass Prosobranchia

The foot in different prosobranchiates may becomes variously modified in relation to different conditions of life.

In several pectinibranchiates the anterior portion of foot is only well differentiated into propodium and some times metapodium equally developed, e.g., *Natica* and *Xenophorus.* In these forms when the animal is in motion it first advances its propodium and then folds the rest of the foot after it. In some forms propodium also forms a siphon on the left side which conduct the water into mantle cavity. The operculum is always carried by the metapodium. Epipodium is usually absent among the pectinibranchiates but vestiges of it can be seen in the form of a pair of anterior pedal tentacles in *Vermetus,* in the form of two nuchal lobes in *Ampullaria.*

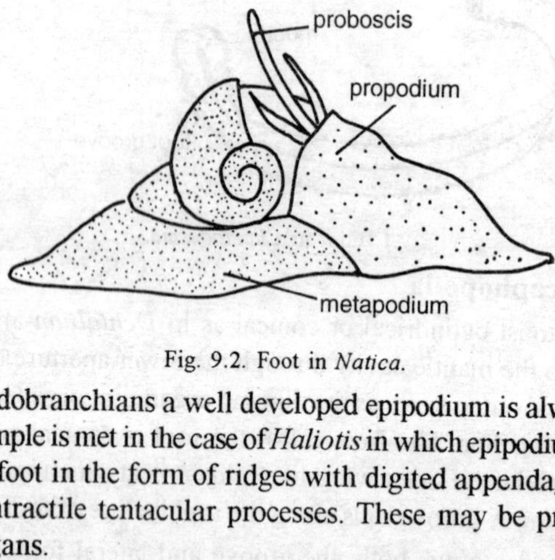

Fig. 9.2. Foot in *Natica.*

In aspidobranchians a well developed epipodium is always present. A good example is met in the case of *Haliotis* in which epipodium surrounds the base of foot in the form of ridges with digited appendages and long sensory contractile tentacular processes. These may be provided with sensory organs.

Broadly speaking, most of the prosobranchiates creep or attach by means of flat sole but in several pelagic pectinibranchiates foot has become especially modified in accordance with swimming mode of life and propodium is changed into a narrow vertical fin and the development of fin can be traced almost step by step starting from *Oxygyrus* proceeding through *Atlanta* and *Pterotrachea.*

(a) *Oxygyrus.* Foot consists of propodium which anteriorly possesses a fin-like outgrowth and a distinct metapodium directed backward like a tail and bearing operculum. The sole is mainly used as sucker.

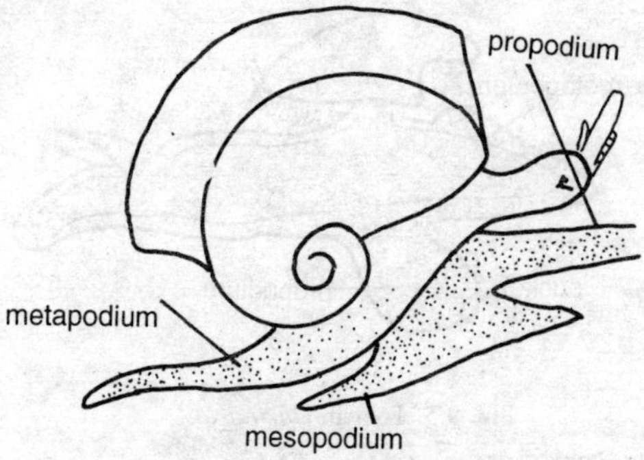

Fig. 9.3. Foot in *Oxygyrus*.

(b) *Atlanta.* The arrangement of foot is similar to that of Oxygyus, but the fin-like out growth of propodium has become its most important part. The comparatively reduced sole or sucker appearing only an appendage to it.

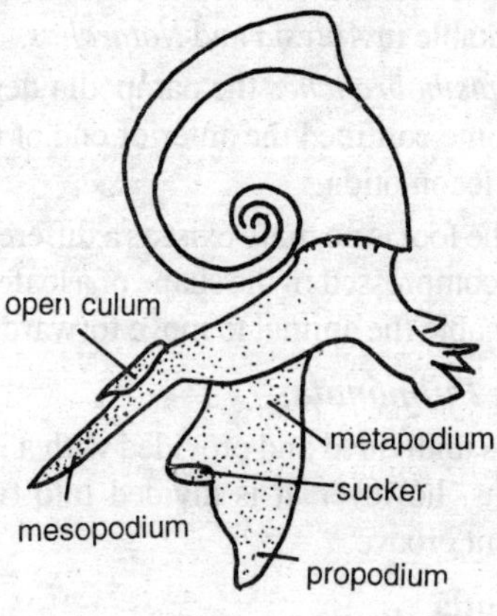

Fig. 9.4. Foot in *Atlanta*.

(c) *Carinaria.* Both, the foot and general external appearance of entire animal, are much changed. The metapodium which here has no operculum appears as a posterior prolongation of body. The fin is much broader and larger and sucker is present just along 1/3 of its free edge.

(d) *Pterotrachea.* The sucker persists only in male, while the metapodium loses the operculum and becomes reduced to a short, filament like a tail.

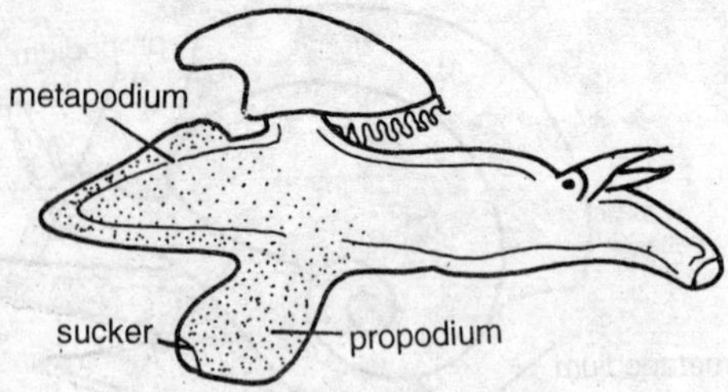

Fig. 9.5. Foot in *Carinaria*.

B. Foot in sub-Class Opisthobranchiata

While there is no division of foot into various parts and among the epipodia are lacking but parapodia are highly developed. These parapodia are lateral lobes of fold-like extensions of the edges of the sole of foot and they are often bent back over the shell so that their edges sometimes touch each other. Thus the shell is completely covered by them. These parapodia may even fuse along their upturn edges at the posterior end. Such a fusion is remarkable in *Aplysia* and *Notarchus.*

Among pelagic *opisthobranches* the parapodia develop into paired fins and foot may become confined the anterior end of the body and two lateral fins are used in locomotion.

In nudibranches the foot is no more exist as a differentiated organ but the body of animals is compressed in the shape of a leaf with sharp dorsal and ventral edges to enable the animal to move forward e.g, *Eolis.*

C. Foot in subclass Pulmonata

The foot is always undivided and provided with a large, flat sole for creeping. In few forms, however, it is divided into two portions by a temporary or permanant groove.

5. Foot in Pelecypoda

Foot, like that of majority of mollusc, is a locomotory organ present as muscular projection from the ventral surface. Usually in this class the foot is painted axe-shaped with sharp edge directed downward and outward. In Protobranchiate for examples *Nucula* and *Yoldia* the foot has a flat, creeping surface and is considered the most primitive foot in the class.

The foot of lamellibranchiates varies according to the mode of locomotion and according to development of byssus. It is secreted by the byssus gland and consists of large number of threads vary in thickness. With the help of these fibres the animal anchors to the foreign bodies.

The complete byssus apparatus consists of:

(a) A cavity in the foot into which the byssus gland opens.

(b) A duct connecting the cavity to exterior.

(c) A groove which runs from the opening of the duct along the ventral edge of foot to its anterior end.

(d) A crescent-shaped or cup-like widening of the groove at its anterior end where the threads are present.

The relation existing between the development of foot and that of byssus apparatus may be given as follows :

(i) The foot is in primitive form with a flat sole and no groove and with a simple invagination but without byssus e.g., *Solemya.*

(ii) From the same foot small lamella arises from the base of the simple invagination and byssus is very slightly developed e.g., *Nucula.*

(iii) Invagination becomes differentiated into a cavity and the byssus and its gland are highly developed e.g., *Mytilus*. In *Arca* and *Mytilus* the old byssus may be shed and a new one formed.

(iv) Many lamellibranches in the adult stage have neither byssus nor byssus gland but the cavity and the duct and even the retractor muscle may be retained. Byssus apparatus may be found into closely related forms without the byssus itself and sometimes with it. In the former case the foot is generally more developed and serves for locomotion e.g., *Trigonia.*

(v) The wedge-like or blade-like foot becomes more strongly developed as a fleshy and extensile organ every trace of byssus and its apparatus are absent e.g., *Unio, Cardium, Tellina* etc. The anterior end of the foot forms a long cylindrical tentacle in *Poromya*, with a swollen free extremity in *Axinus.*

6. Foot in Cephalopoda

The foot in Cephalpoda is represented by arms and siphons. The arms are regarded as the lateral processes of the molluscan foot which have reached beyond the head and united infront and the head is encircled by the foot and the mouth comes to lie in the centre of this. This circle of arms is derivative of the foot which is proved by its anatomical and ontogenic evidences which are as follows:

(i) The arms or tentacles are innervated by the brachial ganglion which is anterior origin of pedal ganglion and it lies under the oesophagus.

(ii) The arms do not occupied in the embryo their definitive possession from the mouth but arise on the ventral surface behind the mouth, and between the mouth and anus in a row found each side. These two rows shift to form a circle of arms in the adult condition.

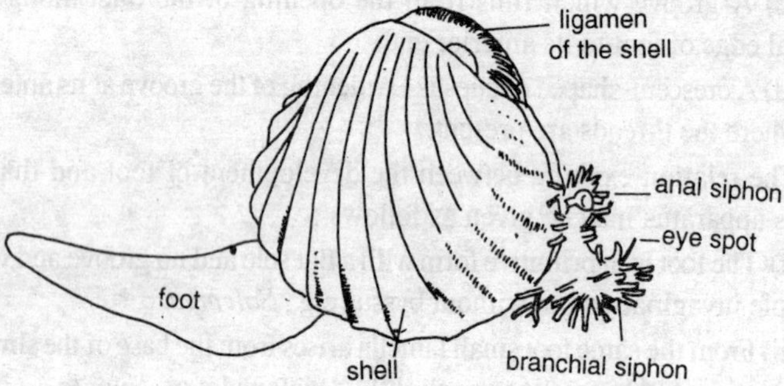

Fig. 9.6. Foot in *Cardium.*

A. Foot in subclass Dibranchiata

In Dibranchiata 8 or 10 such arms are present and each one is provided with two rows of suckess. In *Sepia* more than two rows on ventral side are present. These suckers are stalked in Decapoda and sessile in Octopoda.

The Octopoda has eight arms connected by means of membrane at their bases. In decapods there are ten arms, eight of which correspond to the eight armed octopods but they are not connected by the membrane at their bases. The remaining two arms are longer with swollen ends and their suckers being confined to their free ends. Each of these two arms is situated between 3rd and 4th octopods arm on either side. They lie retracted into special cavities of head and thrown out of their cavities with great force, while catching the food. One, rarely both, of arms is modified to assist in copulation and called as *hectocotylized arm*.

B. Foot in subclass Tetrabranchiata

The head in *Nautilus* bears several tentacles arranged in circle towards the mouth about 90 in females and 60 in males. These tentacles are present on special lobes which may be taken to correspond to arms of dibranchiates while tentacles to the suckers. These tentacles are adhesive and prehensile and are retractile within special tentacular sheaths. In the female there are three tentacular lobes in immediate contact with buccal aperture, these are right, left and ventral-anterior lobes. In males the ventral-anterior lobe is absent instead it bears *Owen's organ*, which is supposed to be olfactory in nature.

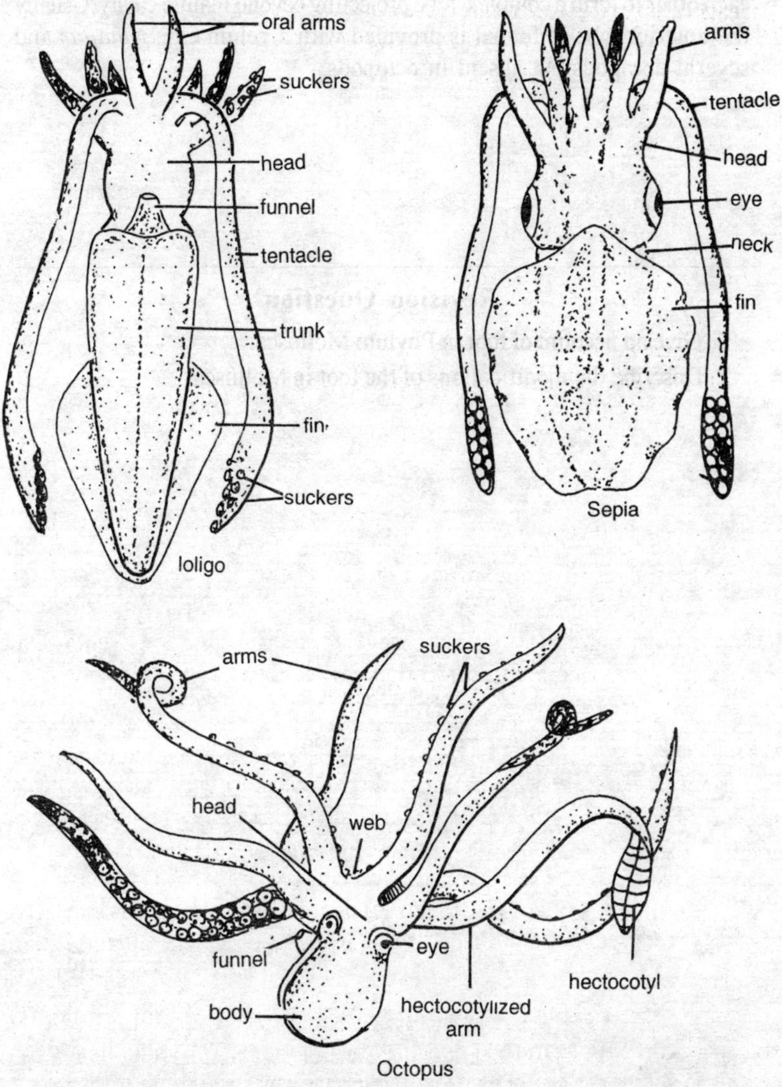

Fig. 9.7. Foot in *Cephalopoda.*

In addition to foot proper the cephalpods possess *epipodia* which are highly modified to form the funnel or siphon. In *Nautilus* the funnel consists of two symmetrical lateral lobes which are enclosed towards each other and overlap sometimes. But in dibranchiate these lobes fuse with each other to form a complete tube projecting beyond mantle cavity. Usually the anterior lobe of funnel is provided with a velum e.g., *Nautilus* and several decapods but absent in octopods.

Revision Question

1. Give an account of foot in Phylum Mollusca.
2. Describe the modifications of the foot in Mollusca.

10

Respiration in Mollusca

The requirement of oxygen presents little difficulty to small organisms. Here the diffusion is adequate to carry out the requirement of oxygen to the cells. It is among the molluscan animals with their relatively large size and covered body that we first meet distinct specialized respiratory structures. Primitively these structures were probably simple projections of the epithelium of mantle but in most present molluscs, these projections called as *ctenidia*. The ctenidia are greatly modified in the direction of increase in surface area and greater vascularity. The second important development among respiratory organs found in connection with the ability of certain gastropods to live permanently on ground. In pulmonates the ctenidia have disappeared and the walls of mantle cavity have folded to form a sac-like structure which is highly vascular. So in molluscs there are three types of respiratory organs.

1. Ctenidia
2. Pulmonary sac or cavity
3. Skin

1. Ctenidia

Ctenidia are the organs of aquatic respiration and are mostly present in aquatic molluscs. The pulmonary chamber or sac is the organ of aerial respiration and is present in terrestrial molluscs. But in certain cases like *Pila*, *Planorbis* etc.both the methods of respiration are employed. Some animals may be regarded as in process of change from an aqueous to terrestrial life and as we see in *Siphonaria*, the ctenidia are reduced to a great extent since the animal comes out of water on the average 22 hours out of 24 hours.

Structure of Ctenidium. The ctenidium has a *central axis*, and one or two rows of *lamellae* or *filaments*. The entire gill is surrounded by cilia. The ctenidium receives venous blood through an afferent branchial vein and after oxygenation, in filaments, it is send back through an efferent branchial vein.

Number of gills. The number of gill vary from 1-80 in different classes of Mollusca.

Type of gills. In molluscs following types of ctenidia are seen:

1. Monopectinate
2. Bipectinate
3. Plicate
4. Holobranchs
5. Mesobranchs
6. Secondary gills.

1. *Monopectinate*. The filaments or lamellae are arranged in a single row on one side of the axis.

2. *Bipectinate*. The gill has two rows of gill filaments on the sides of central axis.

3. *Plicate gills*. These are transverse folds of integument.

4. *Holobranchs*. The ctenidia extend all over the body.

5. *Merobranchs*. When the ctenidia are restricted to a particular area of the body.

6. *Secondary gills*. The true gills are absent. But certain other structures function as gills. They are the following :

(i) *Anal gills*. Gills present around the anus.

(ii) *Cerata*. These are slender vascular bodies located on the dorsal surface.

(iii) *Pallial gills*. Gills arranged in single row around the foot.

In Monoplacophora

In Class *Monoplacophora*, 5 pairs of gills are present in pallial groove. Vestigial gills are supposed to be present on the opposite side of the ctenidial axis. The gills are probably not true gills, more likely they represent prectenidial structure e.g. *Neopilina*.

In Amphineura

In order Placophora (*Chitons*) the ctenidia are arranged in two rows. The number vary from 6-80. According to species when their number is few, the ctenidia are mostly confined to the posterior end. It is not easy to decide, whether the multiple form modification found in *Chiton* is a simple or evolved one.

In the animals of Order *Aplacophora* the foot is absent and the mantle cavity is greatly reduced in the form of a cloacal cavity at the posterior end. In *Chaetoderma* there are two samll, feather-like gills, symmetrically placed on either side of the anus. In *Neomenia* the gills are further reduced and consist of single bunch of filaments lying below the anus. Such gills are restricted to a limited space at the posterior end of the body, and are called *merobranchial type*.

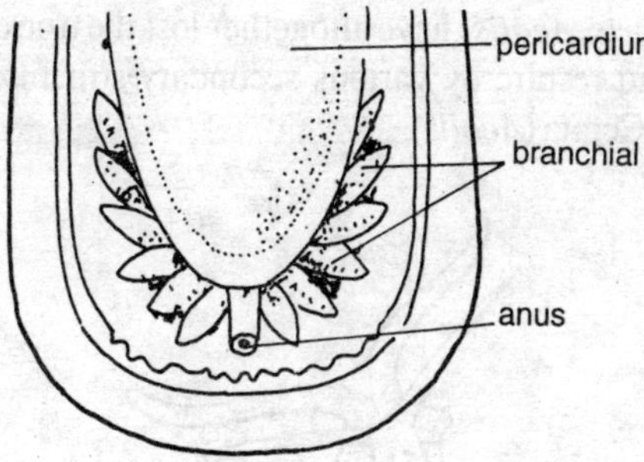

Fig. 10.1. *Chiton.* Showing gills.

In Gastropoda

The gills show a good number of variations started from bipectinate paired structure to a monopectinate single structure due to the assymmetry and in some cases certain secondary structures or adaptive gills are developed which subserve the function of respiration. The gills are shifted in front of the body within the mantle cavity as a result of torsion.

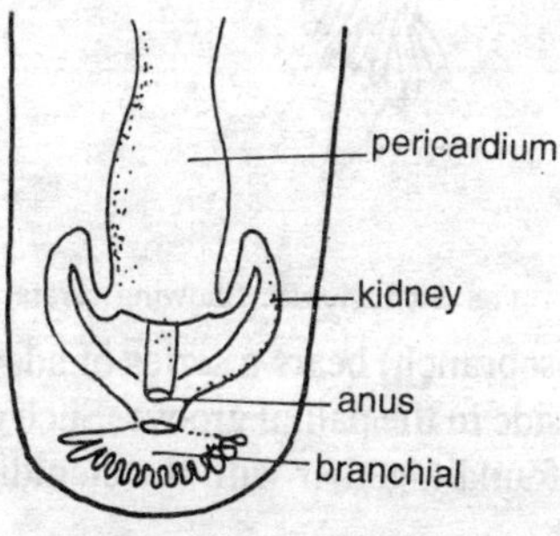

Fig. 10.2. *Neomenia.* Showing gills.

In Prosobranchs symmetrically paired gills occur only in *Haliotis* and *Fissurella*. In all others the asymmetry of the body has caused one of the branchi to become reduced and consequently it is only one branchus usually left in vast majority of prosobranchs. Even in *Haliotis* the right branchius is rather smaller than the left. In aspidobranchs (*Trochus*,

Acmaea etc.) the gill is bipectinate but in pectinibranchs like *Triton*, *Pila* etc. the gill is monopectinate.

In opisthobranchs a true ctenidium only occurs in the *Aplysia*. It lies on the right and usually more or less external being partly covered sometimes quite shallow and sometimes by a fold of membrance. The Nudibranchs (*Doris*, *Aeolis*) have altogether lost the true ctenidium with mantle cavity and respire by various secondary structure such as anal gills (*Doris*) and Cerata (*Aeolis*).

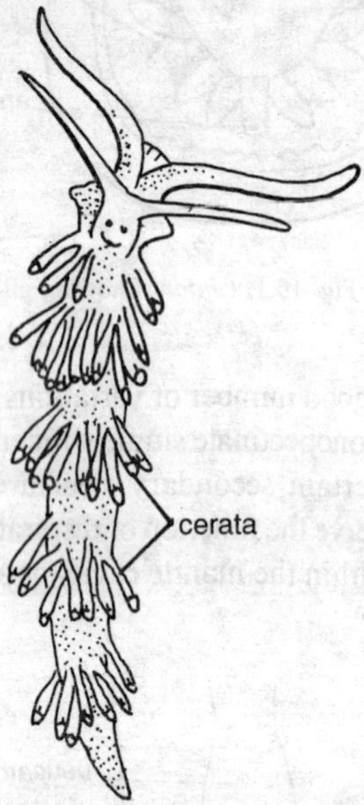

Fig. 10.3. *Aeolis*. Showing cerata.

In *Patella* (Prosobranch) bears a series of adaptive gills occurs in a row on each lateral side in the pallial groove. Such gills are called *Pallial gills*. They may be found together with the ctenidium in *Pneumoderma* (Ospisthobranch).

In Pelecypoda

The ctenidia are two in number right and left, each consisting of a horizontal axis bearing two rows of filaments. They lie one on each side of foot. They lined a chamber known as *branchial chamber*. Leading into this chamber and behind it are two siphons, one of which conducts water to the branchial chamber called *inhalent siphon* and other carries the water out side is called *exhalent siphon*. In many cases like *Mya* both the siphons are exceedingly long, sometimes longer than the whole shell.In some cases the two siphons are free throughout their entire length while in others they become fused. In other genera which do not burrow like *Pecten*, *Mytilus* etc. the siphons are either reduced or altogether absent.

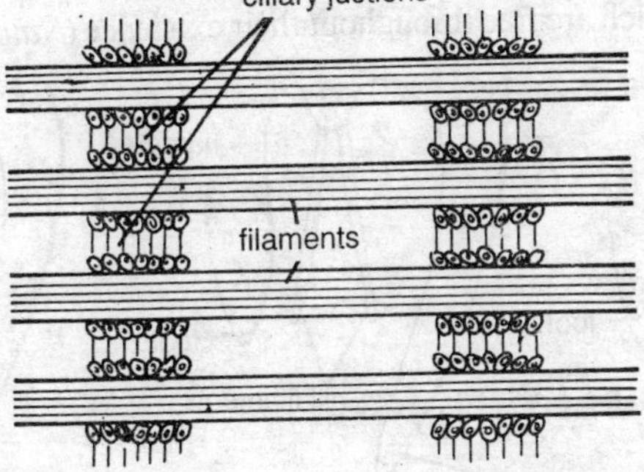

Fig. 10.4. *Mytilus.* Gill filaments magnified.

In Filibranchates, the gill-filaments become elongated and thread-like. In *Arca* each gill filament is bent upon itself to form an elongated 'V'.Thus each gill forms a 'W' in section and consists of two 'V' shaped *demibranchs*. Each demibranch is two lemellae thick, made up of a descending limb and an ascending limb. In *Mytilus* the gills are strengthened by the development of delicate non-vascular bar or inter-lamellar junctions between the two limbs of each filament. Thus the single ctenidium of *Nucula* is replaced by two plate-like laminae in *Mytilus*. Each lamina is made up of an outer and an inner *lamina*.

In *Pecten* and *Ostrea*, the gills have a greater cohesion than in filibranchs and the reflected distal ends of filaments become fused laterally with the mantle.

In Eulamellibranciates (*Unio*, *Anodonta*) the ciliary interfilamentar junctions are replaced by vascular cross cinnections with narrow openings between them called *ostia*. The interlamellar junctions also become large and vascular, so that the inter-lamellar space is partitioned into vertical water tubes. The incomming water enters the water tubes through these ostia.

In Septibranchiates (*Poromya*, *Cuspidaria* etc.) the gills are represented by a muscular septem which divides the mantle cavity into the dorsal and ventral parts. This septum is perforated at certain places, which allow the communication between dorsal and ventral chambers.

In Cephalopoda

The ctenidia are situated in the pallial cavity on either side of the visceral mass. In Dibranchiata, there are only one pair of gills. A gill is composed of lamellae whose number varies in different forms. Each lamella is thrown into transverse folds which are inturn folded so that the respiratory surface is largely increased. The gills are attached dorsally to the mantle (*Sepia*, *Loligo* and *Octopus*). In *Tetrabranchiata* there are two pairs of gills which are free throughout their existence (*Nautilus*).

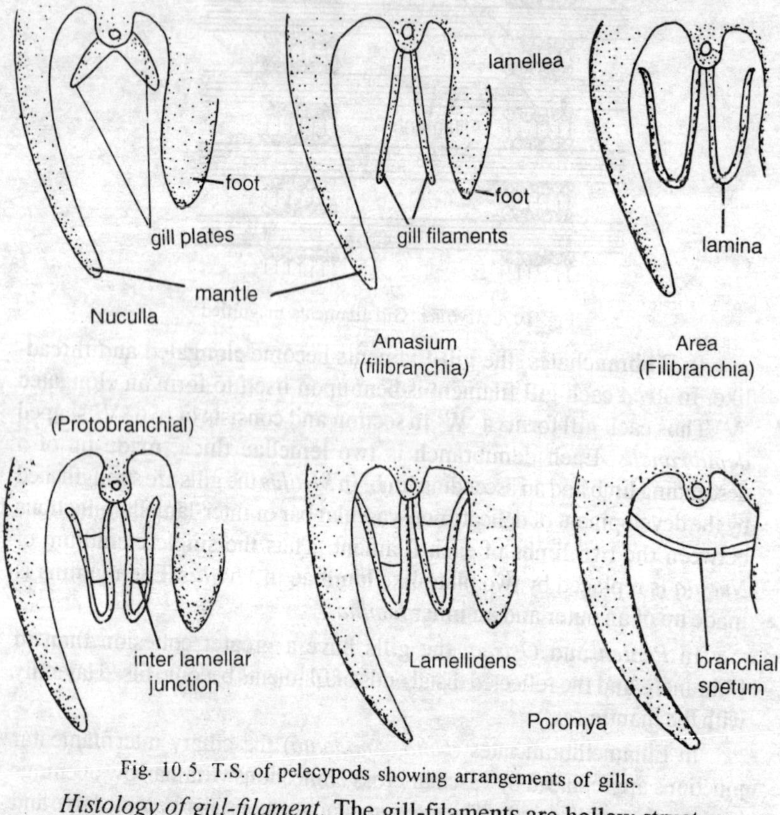

Fig. 10.5. T.S. of pelecypods showing arrangements of gills.

Histology of gill-filament. The gill-filaments are hollow structures and each filament has two sides. One side facing towards the mantle cavity and other towards the inter-lamellar cavity. The gill filaments are provided with chitinouss rods to give a sort of rigidity to the gill filaments. They are covered with ciliated epithelium. The epithelial cells are larger on the side facing mantle cavity and secrete mucous. The presence and arrangement of cilia is remarkable, on the sides are lateral cilia, on the outer ridge-like faces of filaments are frontal cilia and those lying in between are latero-frontal cilia.

Pulmonary Sac or Cavity

It is an organ for aerial respiration and is found in land molluscs, mostly in pulmonates and some amphibious prosobranchia. When we use term 'lung' it must be remembered that this organ in mollusc does not correspond morphologically with spongy, cellular lungs of vertebrates. It simply performs the same function. The lung in mollusc is a cavity or pouch lined by blood vessels which are disposed over its surface in various ways forming network.

In adaptation to terrestrial mode of life the pulmonary respiration is resulted. Therefore it is found in vary different groups of *Gastropoda*, but different stages of evolution are best studied in *Streptoneura*. In this group certain aquatic forms, though they possessed ctenidia, required the habit of living for short or longer time out of water. Consequently certain modifications have taken place in the internal structure of mantle. So that in certain forms such as *Ampullaria* and *Pila*, both ctenidia and pulmonary sac are present and the mantle cavity is divided into right *pulmonary chamber* and left *branchial chamber* by an incomplete septum. In others the ctenidia are totally absent and the pulmonary chamber is fully developed (*Gadinia*).

Outer skin

Respiration by moist-skin is the simplest method of respiration in molluscs. In certain aquatic (*Scaphopoda* and terrestrial *Opisthobranchiata*), terrestrial (Vaginulidae) and parasitic (*Ontodax* and *Entoconcha*) neither ctenidia nor localized respiratory organs are present. Respiration occurs through entire body surface or soft-skin of the body. In *Onchidium* respiration is, however, carried by skin as well as by dorsal papillae present on skin. Skin functions as respiratory organ even when ctenidia, as well as pulmonary cavity is present.

11

Torsion and Detorsion among Gastropods

In all the gastropods, with coiled shells, the body is asymmetrical with the mantle cavity lying anteriorly, and the coiling of visceral hump and shell are directed posteriorly. But in the development of these forms from the larva, the mantle cavity first makes its appearance behind the visceral hump and at a particular stage the visceral hump rotates in a counter-clockwise direction through an angle of 180^0 on the rest of body. This is known as *torsion*. Thus torsion may be defined as, "the rotation of the visceral organs to an angle of 180^0 in the anticlokwise direction". It occurs in the larva of gastropods and converts the symmetrical larva into an asymmentricl adult.

PROCESS OF TORSION

It occurs in two stages:

1. Ventral flexure
2. Lateral torsion

1. Ventral flexure

It takes place in the anterio-posterior to the sagittal plane about the transverse axis situated at right angle to the main anterio-posterior axis of the animal. It results in the looping of alimentary canal. As a consequence of this flexure the visceral mass, the shell which was originally saucer shaped, becomes more or less pointed cone and during this process the visceral sac finally assumes spiral coiling.

2. Lateral torsion

The ventral flexure is soon followed by lateral torsion and dorsal shell becomes ventral in position. This is brough about by unequal growth on two sides. Generally the organs present on the left side grow normally and of the right side has retarded development. The organs from left side shift towards right. The mantle cavity and pallial complex are shifted towards the right side and becomes anterior in position.

EFFECTS OF TORSION

Present gastropods are asymmetrical showing twisting of nervous system and asymmentry of internal organs. The peculiarities of organisation of gastropods, due to torsion, were firs trealized by *Spenge* (1881). The effects of torsion are not equally evident in all gastropods. Following changes are seen after torsion :

1. Before torsion the ctenidia, anus, renal aperture point backward and the auricles lie behind the ventricle. After torsion ctenidia. project forward, anus lies on right side and auricles lie anterior to the ventricle. The original posterior face of the visceral sac becomes the anterior face.

2. Nervous system becomes twisted into figure of 8 and most of the ganglia concentrate in the anterior part of the body. The pleural and pedal ganglia fuse together and lie immediately beneath the cerebral ganglia forming a ring around buccal mass.

3. The alimentary canal which was straight from mouth to anus, is thrown into a loop.

4. The mantle cavity was primitively posterior in position but after torsion the mantle cavity opens just behind the head and its associated parts are shifted forwards.

How torsion occurs?

Thomson (1958) recognized seven ways in which torsion occurs:

1. Complete or 180^0 rotation is achieved by muscle contraction alone. It is known only for *Acmaea*.

2. 180^0 rotation achieved in two stages, first 90^0 movement by contraction of larval retractor muscle and later 90^0 rotation by different grwoth e.g. *Patella*, *Haliotes*.

3. 180^0 rotation by differential grwoth process alone e.g. *Vivapara*.

4. Rotation by differetial grwoth process, with anus coming to a position appropriate to the adult state as in *Aplysia*

5. Torsion no longer recognisable as a movement of viscero-pallium e.g. *Adalaria*.

6. The visceral mass and shell were dorsal now becomes ventral by torsion.

7. The original symmetry which the animal possessed before torsion now lost. The ctenidium, osphradium, auricle and kidneys more developed on one side then other. In some specialized froms (*Pila*) the ctenidium, kidneys and aurieles etc, are present on right side and disappear from left side, but in few primitive forms like *Patella* the symmetry is prevented by

the presence of two gills, aurciles and kidneys but the right kidney serves as gonoduct.

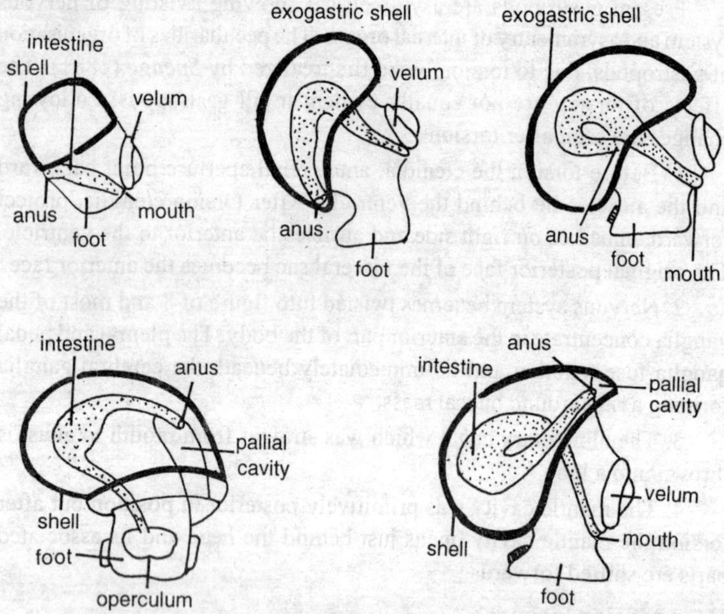

Fig. 11.1. Five successive stages showing development of a gastropod to explain torsion. A—Early embryo without ventral flexture. B—Embryo with an exogastric shell in a lateral view. C— Fig. showing 90° of a lateral anticlockwise torsion with endogastric shell while mantle and anus moving towards right side. D—90° torsion stage (Posteriorly). E—Adult form with complete torsion (180°) in lateral view.

THEORIES OF TORSION

There are various theories to explain the cause of torsion. A few are described here :

1. Torsion occures in the embryo due to antogonism between the growth of foot and that of shell (*Banton*. 1919).

Due to development of one piece protective shell, which encloses the soft parts inside it, opens on one side through which the head and gills are to protrude out and for this the head as well as the posterior extrimity of the canal have to be brought about on the same opening of shell. This requirement shifted the ctenidia and mantle cacity to come towards the head end. This is also existed by the extensive development of the foot. The development of foot results in the shifting apart of mouth and anus which is brought about near each other during larval stage by the ventral flexure and lateral flexure. Therefore, the foot is necessary for the phenomenon of torsion.

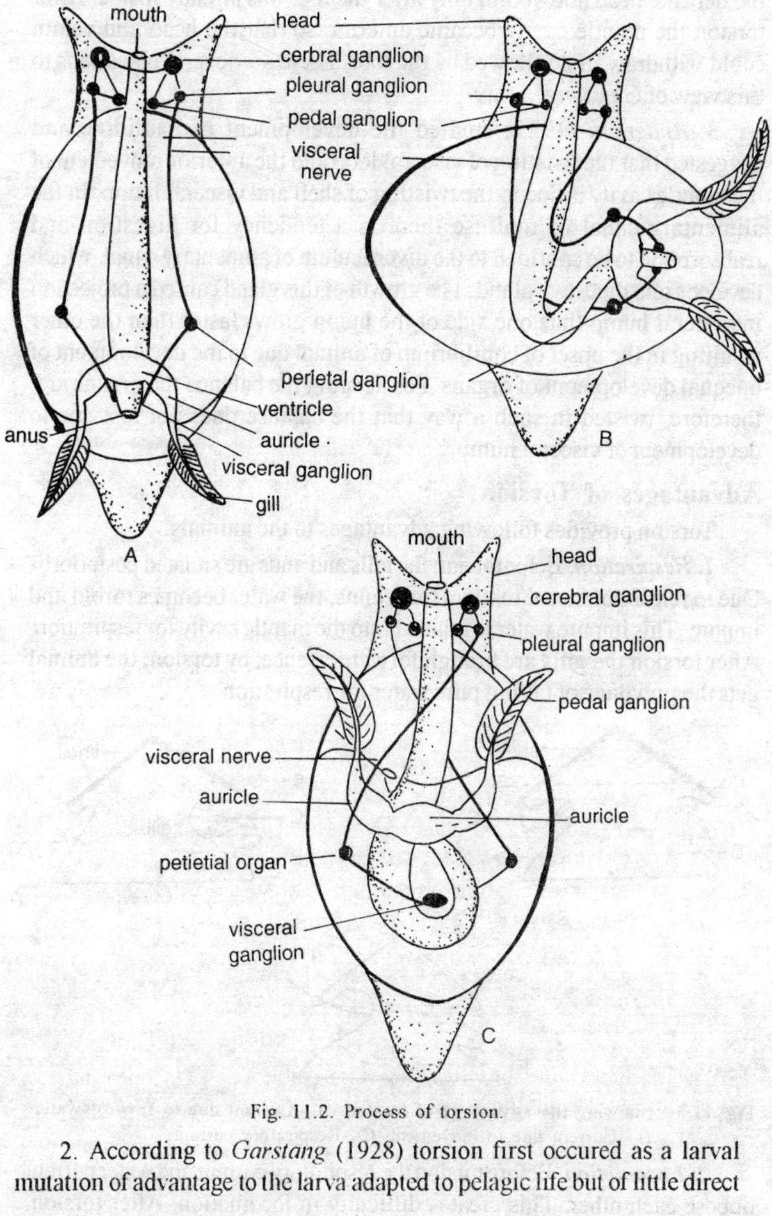

Fig. 11.2. Process of torsion.

2. According to *Garstang* (1928) torsion first occured as a larval mutation of advantage to the larva adapted to pelagic life but of little direct

use to the adult. Before torsion, the untwisted swimming larva fell an easy victim to its predators because the posterior mantle cavity could receive the delicate head and velum only after the foot was already inside. After torsion the mantle cavity became anterior, so that the head and velum could withdraw first followed by the foot. There are several objections to this view of *Garstang*.

3. *Anderson* (1925) studied the development of Paulidine and suggested that the twisting of visceral loop and the anterior movement of the mantle cavity is due to the twisting of shell and visceral hump. In the alimentary canal of mollusc there is a tendency for digestion and reabsorption to be confined to the diverticulum of alimentary canal, which develops into digestive gland. The growth of this gland caused a projection in visceral hump thus one side of the hump grows faster than the other resulting in the upset of equilibrium of animal due to the development of unequal development of organs. For keeping the balance the organs are, therefore, twisted in such a way that the balance does not lost due to development of visceral hump.

Advantages of Torsion

Torsion provides following advantages to the animals:

1. Respiration. Befor torsion the gills and anus are situated postoriorly. Due to release of faecal matters from anus, the water becomes turbid and impure. This impure water is inhaled into the mantle cavity for respiration. After torsion the gills are brough forward. Hence, by torsion, the animal gets the previlage of taking pure water for respiration.

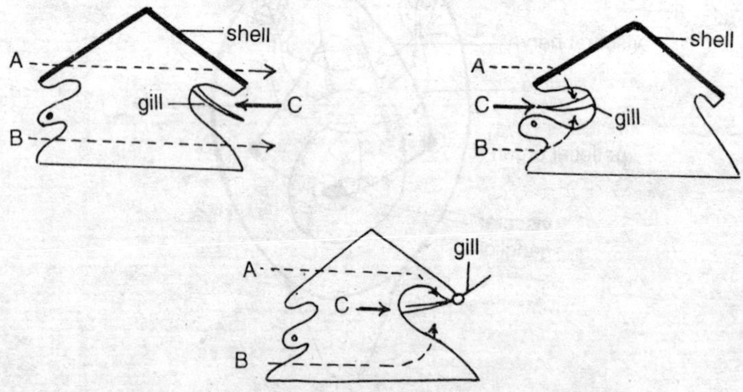

Fig. 11.3. Showing the advantage of torsion. A—Current due to flow of water. B—Current due to movement. C—Respiratory current.

2. Locomotion. Before torsion the respiratory current and water current oppose each other. This creates difficulty in locomotion. After torsion.

gills are placed forwards and the respiratory water is taken from ahead. So the respiratory current concides with the water current. This does not produce any hinderance to locomotion.

*3. **Protection.*** Before torsion the foot is withdrawn into the body first. After foot, the head and ctenidia are taken in. But after torsion the head and ctenidia are taken in first and lastly the foot is withdrawn. This gives protection from enemies.

Disadvantages

Due to torsion following disadvantages are seen:

*1. **Sanitation.*** Torsion brings the anus close to the mouth, thus the faecal matters released from the anus gives rise to the sanitation problems.

*2. **Respiration.*** The faecal matter is released forward. It makes the water lying ahead impure which is inhaled for respiration.

DETORSION

Changes occuring in torsion are reversible and these change are known as *detorsion*. The anus and other organs of pallial complex are carried back to its original position, which are posterior, by detorsion. Thus the lost symmetry reappears to forming symmetrical animal.

In least specialized opisthobranchiats and pulmonates, the torsion is not complete and the pallial aparture is carried only to right of lateral side not to posterior end but in most forms like *Aplysia* the anus and the pallial cavity are moved back to posterior extremity of body. The tendency of detorsion may be observed in exceptional cases among the Streptoneura but it is marked in the whole group of Euthyneura includidng opisthobranchiates and pulmonates.

EFFECTS OF DETORSION

Effects of detortion can be noted as follows:

1. External morphology

Detorsion is always associated with the loss of shell and thus the libration of gills from their enclosing case. Once the shell is lost the gills become exposed to external current and the anterior position of gills is great of advantage.

The degree of detorsion can be correlated with the extent of which the shell is reduced (Pteropoda and Tectibranchs).

2. Mantle and Pallial complex

Detorsion has created condition in stand. According to analysis of *Perrier* and *Fisher* (1911) the left side of mantle roof, taking various structures with it, has been showed to the right in such a way as to becomes part of the mantle floor, fusing with the neck of animal. Thus some structure that are borne of the mantle roof in prosobranchs come to lie in the mantle in opisthobranchs. Also in the detorsion process, a right posterior part of the mantle become detached from the shell and hungs loose as an infra-pallial lobe that extands posteriorly and may attain considerable size as in *Philine* and other Cephalaspids. In *Akera* the posterior edge of the infra pallial lobe bears a long tenacle as also in *Aglaja*. In some cephalaspids, notable the Actenoidae, the mantle is drawn out posteriorly into a conspecuous caecum which bears internally a dorsal and a ventral ciliated tract concerned in maintaning a water current through the mantle cavity. The caecum may project freely, often spirally round, or may be fused to the surface of visceral mass. The mantle of opisthobranchs generally lacks siphons but in *Aplysia* and other anaspids the anus is located on a projecting spout of the posterio-lateral corner of mantle called by *Haffmann* (1932-1939) the *cloacal siphon*, a site of an out going current. A true mantle sipon may be substituted by temporary rolling into a tube by the posterior margin of the cephalic shield as in *Ringicula* and *Gastropteron* as already noted.

3. On alimentary canal

The alimentary canal before detorsion and after torsion makes coil but after detorsion it again comes to its orginal positon and becomes straight as it was before torsion.This gives the symmetry in animal.

4. On respiratory system

Due to detorsion the shell is lost and the respiratory organ such as gill or ctenidia are librated. The gills become directly exposed to external water current. The anterior position of gills is no longer so great and advantage and that is why the gills are situated in the posterior region on either side of the anus in detorted forms.

5. On nervous system

The pulmonate central nervous system is of detorted type with the ganglia concentrated into a ring around the alimentary canal. Usually this circum-enteric ring is located behind pharynx so that the oesophagus and the ducts of the foregut glands must pass through it, but in a few cases it encircles the pharynx anteriorly. The detorsion is brought about by the migration of the intestinal and visceral ganglia forward to contact with the pleural ganglia. Thus eliminating the pleuro-visecral connectives. These connectives are still evident in *Chilina* and traces of them remain in some other primitive pulmonates (Otina, *Ellobium*). Not only detorsion results in the formation of symmetrical nervous system but an advanced concentrate type of nervous system. It may not be a direct effect of detorsion but it follows the sequence of symmetrical development of nervous system.

6. Effect of detorsion on development

The new idea put forward by *Thompson* (1958,1962) about detorsion is against the whole concept of torsion and detorsion in which we believe in the views like of *Pelseneer* (1906-1911). *Pelseneer* stated that in detorted forms the primitive torsion is manifested in the course of development in that in the larval mantle cavity is anterior and dorsal. But this was denied by *Thompson* (1958-62) who noted no indications of torsion in the larvae of two nudibranchs whose development studied by *Pelseneer* nor in other nudibranch larvae examined. It appears that in opistbobranch development there is generally no recapitulation of torsion, but the organs are formed in their definitive sites. However, a posterior displacement of the anal complex is not uncommon. *Thompson* himself noted that at metamorphosis of *Adalaria* the anal complex moves back along the right side of its definative posterior position, thus reversing torsion. During the development of the onchidacean *Onchidella* (*Fretter*, 1943) the right side rotates to a dorsal and the mantle cavity with its contained openings (anus and nephridiopore) migrates back along the right side. In the *Pleurobranchid*, *Berthellina citrina* (*Gohar* ard *Abul-Ela* 1957) the right side rotates to a dorsal position, the dorsal side shifts to the left, and the anus is carried back. A posterior displacement of the anus also occurs during the development of nudibranch *Eolidina curia* (*Tardy*, 1962). These changes may be regarded as detorsional since the main effect of torsion is to bring the mantle cavity and contained complex to an anterior dorsal position. Such a position of the mande cavity does not seem to occur during the development of opisthobranchs. In nudibranch veligers the mantle cavity is generally located lateroventrally on the right side (*Thompson*, 1962).

12

Nervous System in Mollusca

The molluscs prossesses a well developed nervous system which usually consists of a number of ganglia which are linked by commissures and sending out thread-like nerves which ramify into the various organs. The pattern of nervous system varies greatly in different groups of Mollusca, from extreme complex condition to a simplest form in which ganglia are almost entirely absent.

I. Nervous System in Monoplacophora

The nervous system consists of a *circum-oral nerve ring* or *cerebral commisssure* surrounding the mouth. From it arises two pairs of nerves in posterior region, the inner *pedal cords* and outer *visceral cords*, which join posteriorly to form loop. The pedal and visceral cords are connected by 10 pairs of *transverse connectives*. Nerves extend from the cords to the gills.

II. Nervous System in Amphineura

The nervous system is of primitive type. The nervous system consists of *cerebral ganglia* which are placed above the mouth on either side.

The *pedal ganglia* are situated on either side below the oesophagus or at the base of foot. The position of *pleural ganglia* varies considerably but always below the oesophagus and slightly above the pedal ganglia.

In chitons there are four nerve cords, parallel to one another and are present nearly entire length of the mass. The exterior part represent the *pallial cords* and the inner, the *pedal cords* are united by *transverse connectives*. The two pallial cords are united posteriorly by a thick *supra-renal commissure*. Ganglia are lacking or at least poorly developed.

In *Neomenia* there is a distinct *cerebral ganglia* formed by the massing together of the two nerve cords on each side.

III. Nervous system in Scyphopoda

The nervous system is symmetrical. A pair of each *cerebral* and *pleural* ganglia are present close together on the dorsal side of oesophagus.

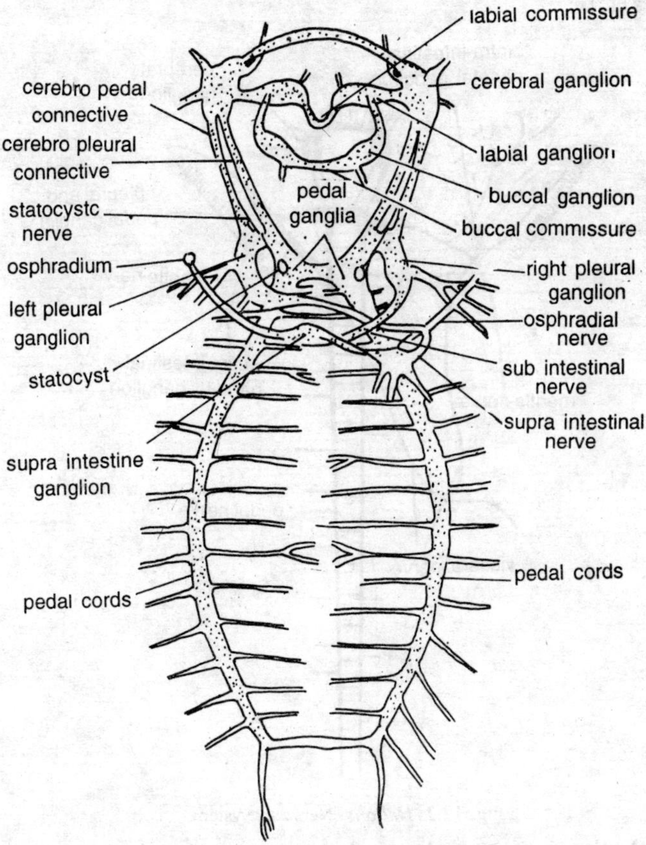

Fig. 12.1. *Patella.* Nervous System.

A pair of *pedal ganglia* lie in foot and united to the cerebral and pleural ganglia. Lying on either side of rectum are *visceral ganglia*, united by *pleuro-visceral connectives* to the pleural ganglia.

IV. Nervous System in Gastropoda

With the exception of endoparasitic Endochonchidae all gastropods possess a well developed nervous system in which *cerebral*, *pedal* and *visceral nerve centres* with their respective connectives and commissures are present. The special character of gastropod nervous system is asymmetry of visceral centres and of nerves arising from them. The condition of nervous system varies greatly in different groups ranging from primitive nervous system of considerable simplicity (*Patella* and

Haliotis) to the concentrated highly evolved nervous system (*Lymnaea*, *Tethyes* etc).

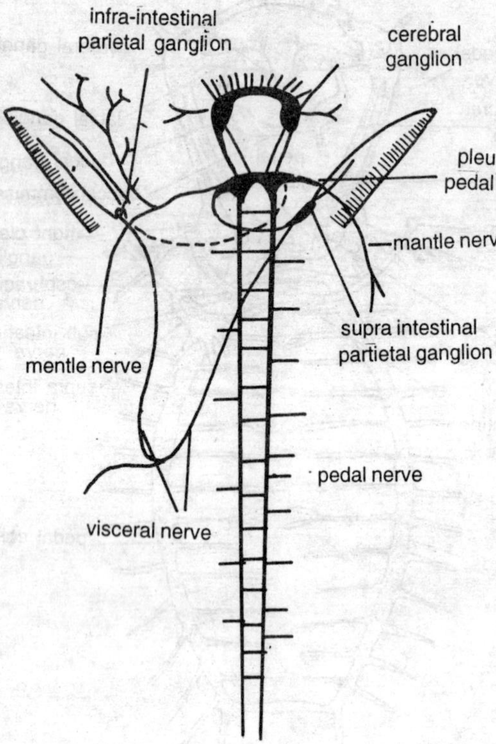

Fig. 12.2. *Haliotis*. Nervous System.

Hypothetical general plane of nervous system in gastropods

The nervous system of gastropods is distinctly ganglionated and somewhat complex. It consists of a pair of adjascent *cerebral ganglia* which give rise nerves to the eyes, tentacles and a pair of *buccal ganglia*. The buccal ganglia innervate the muscles of radula and other structures of buccal mass. From cerebral ganglia, a pair of *pedal cords* arise posteriorly, which extend to a pair of *pedalganglia* located in the foot and supplies the muscles of foot. A fine pair of nerves, the *visceral nerves*, arising from cerebral ganglia and extend posteriorly until they terminate in a *visceral ganglion* located in the visceral mass. Along the course of these nerves are located two pairs of ganglia. Located close to cerebral ganglia are the *pleural ganglia* and more distinct are *parietal ganglia* which supply nerves to gills, osphradium, and mantle.

Evolution of nervous systern in Gastropods. Three tendencies are marked in the evolution of gastropods:

1. The entire nervous system has evolved along two definite lines due to torsion, in one type of nervous system the loop formed by the visceral nerves is persisted over itself forming a figure of 8 i.e. streptoneuran system present in Prosobranchiates. While in other case, the loop remains untwisted i.e. euthyneuran type mostly present in Opisthobranchiates and Pulmonates.

2. Secondly, the nervous system in gastropods shows a marked tendency towards the concentration of various ganglia. In certain pulmonates all the ganglia form a ring around the oesophagus, while finally in *Tethys*, the ganglia of one side fuse with each other.

3. Thirdly, the long pedal cords persist in the primitive cases of *Patella* and *Haliotis* are lost during the evolution and finally the pedal ganglia come to lie near the other ganglia without having long pedal cords.

Types of nervous System on the basis of cerebropleural connectives. three categories have been recognized:

1. Hypoathroid. In this case the pleural the pedal ganglia are quire distinct and the pleural ganglia lie in intimate contact of the anterior part of pedal ganglia and the result is that the cerebropleural and cerebropedal connectives are quite long and nearly equal to one another e.g., *Patella*.

2. Dysathroid. The pleural ganglia, in this case, move toward the cerebral ganglia, with the result that cerebro-pedal connectives are longer in comparision to those of cerebro-pleural connectives e.g. *Ampularia*.

3. Epiathroid. In this case pleural ganglia fuse with cerebral ganglia and the result is that cerebro-pedal connectives are quite long while cerebro-pleural connectives are lost. e.g. various pectenibranchiates.

Nervous system in subclass Streptoneura. The visceral commissures are in the form of 8.

Nervous system in order Aspidobranchiata. Here the concentration of ganglia is little. In most primitive streptoneuran, the cerebral ganglia are placed at the sides of oesophagus apart from each other. They are connected by a cerebral commissure. The pedal ganglia give out too long pedal cords extending along the ventral side united to transverse connective. The pleural ganglia are approximated to the pedal ganglia. Sometimes an *infra* or *labial commissure* is also present. e.g. *Patella*, *Haliotis* etc.

Nervous system in order Pectinibranchiata. The nervous system is somewhat concentrated without a labial commissure except in *Paludina* and *Ampullaria*. The nerve collar is situated behind the buccal mass except in *Ampullaria*. The number of visceral ganglia varies from 1-3. The pedal cords and transverse commissures are absent except in *Palud ina* and *Cyclophorus*. Progressive zygoneury (supra or sub-intestinal ganglion from which all the mantle nerves of that side appear to be given out) is present in majority of cases. It takes place on right side. In other pectinibranchiates sometimes dialoneury (mentle is supplied by nerves of pleural ganglion of same side as well as by a nerve from supra-intestinal ganglion of same side) is present. In *Cassidaria* zygoneury as well as dialoneury is present. In *Natica,* where the anterior part of foot is strongly developed, a pre-pe ganglion becomes differentiated from the pedal ganglion.

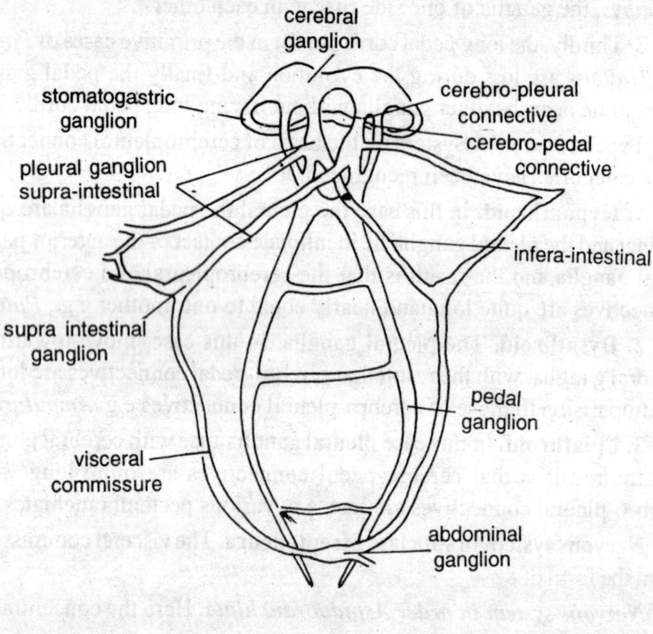

Fig. 12.3. *Paludina*. Nervous System.

Nervous system in sub-class Euthyneura. The nervous system, in which all the typical ganglia are present, is characterized by the following peculiarities :

1. By the absence of streptoneury except in *Acteon*.
2. By the marked tendency towards the concentration along the posterior end of pharynx.

The animals of this group are symmetrical forms having untwisted pleuro-visceral connectives due to detorsion. The external opening and mantle cavity are shifted from anterior to the posterior side and thus regain symmetry secondarily. The ganglia tend to become in a ring-like form united by short connectives and commissures. eg. *Aplysia.*

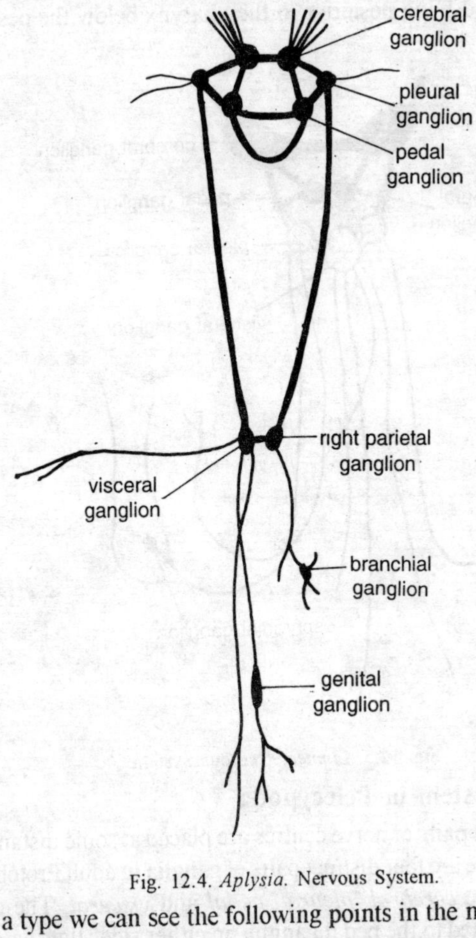

Fig. 12.4. *Aplysia.* Nervous System.

As a type we can see the following points in the nervous system of *Bulla*

1. Pleural ganglia have shifted till they lie close to cerebral ganglia resulting in the shortening of cerebro-pleural connectives.
2. Three visceral ganglia.
3. Parapodia are innervated by pedal ganglia.

Nervous System in sub-class Pulmonata. All typical ganglia are present here. They are put together immediately behind the pharynx and there is present a peri-oesophageal complex into which even the parietal and visceral ganglia have been drawn. The cerebral ganglia lie close on dorsal side, while all other ganglia lie ventrally and the cerebro-pedal and cerebro-plcural connectives are easily distinguishable. The buccal ganglia are always found lying posterior to the pharynx below the oesophagus e.g. *Lymnaea*.

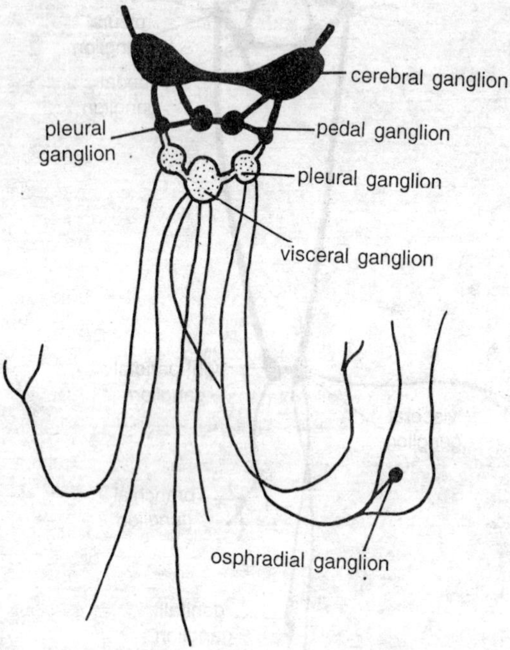

Fig. 12.5. *Limnea*. Nervous System.

V. Nervous System in Pelecypoda

The different pairs of nerve centres are placed at some distance from one another. There are four distinct pairs of ganglia in adult Protobranchs. These ganglia are *cerebral*, *pleural*, *pedal* and *visceral*. The cerebral ganglia are attached to the pedal ganglia on either sides thus having two pedal connectives on each side i.e. cerebro-pedal and pleuro-pedal as in *Solenomya* and *Nucula*.

In all the pelecypods the cerebro-pleural centres are connected with the pedal and visceral centres, therefore,two types of connectives are present the *pedal* and *visceral connectives*. In *Mytilus* the visceral connective is fused for a certain distance with pedal connectives.

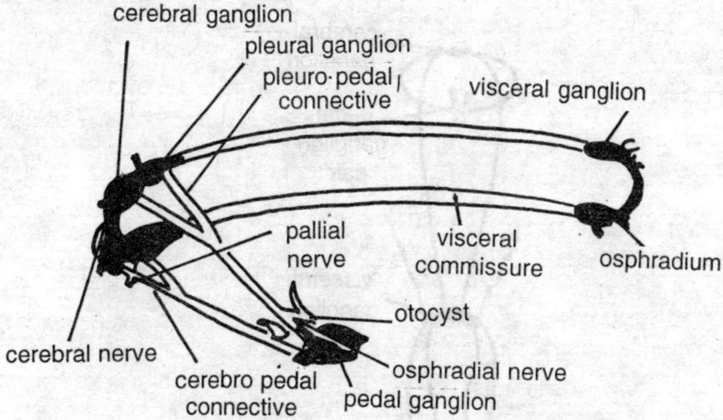

Fig. 12.6. *Nucula.* Nervous System.

In *Protobranchia* and *Venus* the cerebro-pleural ganglia are attached to one another but in all other group they are apart fron each other. The pedal ganglia, situated in the foot, are attached to one another. When the foot is degenerated they also become very much reduced as in *Teredo*, *Pecten* etc.

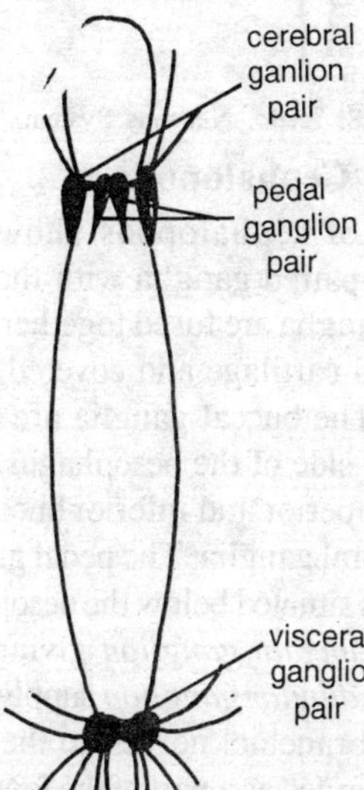

Fig. 12.7. *Pecten.* Nervous System.

The visceral ganglia are situated some distance behind the posterior adductor muscle but in some cases they are found on the ventral side of the posterior adductor muscle except in *Pholas*, *Teredo* etc.

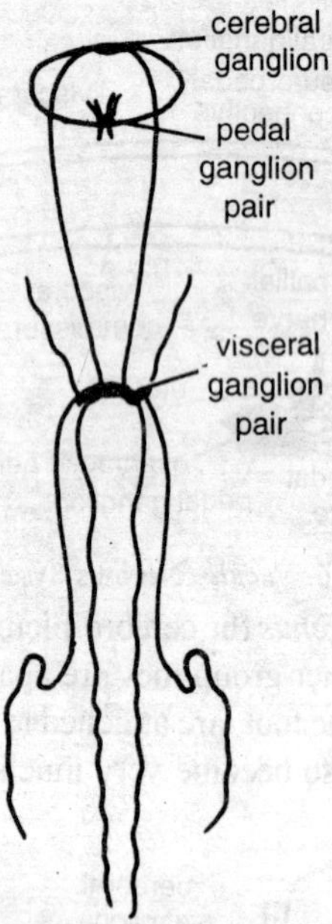

Fig. 12.8. *Terdo.* Nervous System.

VI. Nervous System in Cephalopoda

The nervous system of cephalopods shows a high grade of development. It consists of paired ganglia with their commissures and connectives. The cerebral ganglia are fused together into a rounded mass and protected by the cranial cartilage and covered over anteriorly by a strong fibrous membrane. The buccal ganglia are situated close to the buccal mass on the anterior side of the oesophagus. The buccal ganglia may be differentiated into superior and inferior buccal masses giving the appearence of separate cerebral ganglia. The pedal ganglion is in the form of an united ganglionic mass situated below the oesophagus and is divided into an *anterior lobe* or *branchial ganglion* giving nerves to the arms and a *posterior lobe* or *infundibular ganglion* supplying the funnel. From the branchial ganglion ten branchial nerves, to the arms, are given off, therefore, these arms are regarded as a part of the foot. Branchial ganglion is also connected to the *superior buccal ganglia* by paired *branchio-buccal commissures* and to the *cerebral ganglia* by *cerebrobranchial commissures.*

A pair of *pleuro-visceral ganglia* in the fused state are found situated posterior to the infundibular ganglia. From this ganglia are given off two stout *pallial nerves* to the mantle, where each dilates to form a large

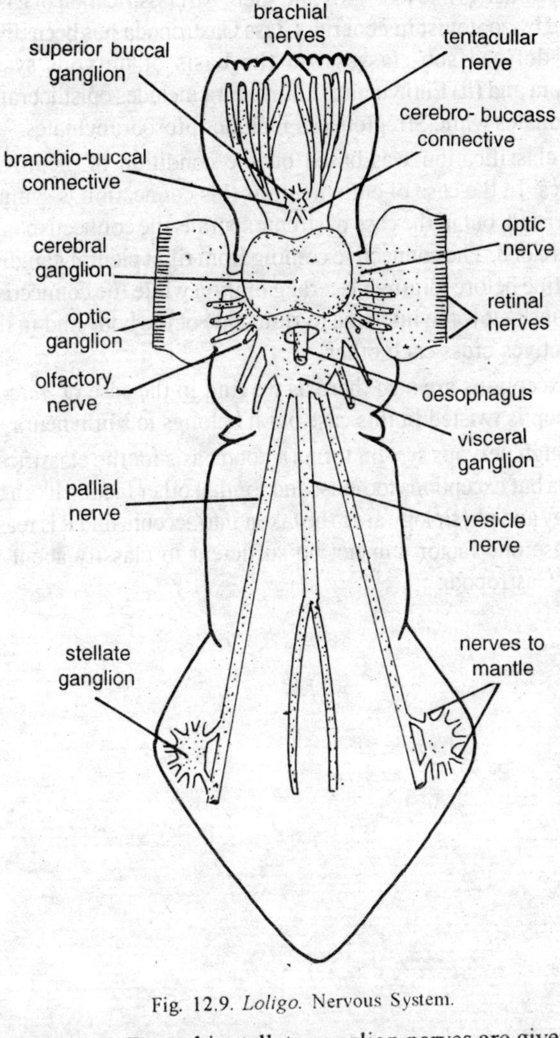

Fig. 12.9. *Loligo*. Nervous System.

stellate ganglion. From this stellate ganglion nerves are given off to the mantle. From the pleuro-visceral ganglia originate paired visceral nerves which supply internal organs (i) from a visceral loop from which springs a pair of branchial nerves, (ii) innervating the gills and each of which expands to form a *branchial ganglion*, (iii) at the base of gill.

Importance of nervous system in Classification

The nervous system of molluscs is also the basis for its classication like other characters like foot, gills and shell. No classification of gastropods is accepted by zoologist in general. Class Gastropoda has been divided in two well defined sub-classess on the basis of nervous system (i) Streptoneura and (ii) Euthyneura. Euthyneura includes opisthobranchiates and pulmonates while Streptoneura includes prosobranchiates.

This classification was based on the condition of pleuro-visceral connectives. In the case of euthyneurans this connection is symmetrical and not twisted, but in the case of streptoneurans the connective is twisted in the figure of 8. The connective coming from right pleural ganglion runs over intestine before joining visceral ganglion while the connective from left side runs below the intestine to right side of the body and in this way the connectives cross each other.

But exceptions are also there as we find in the case of *Acteon*. The visceral loop is twisted in this case but it belongs to Euthyneura.

Although nervous system forms a good basis for the classification of Gastropoda but exceptions too exist and for that other factors like anatomy, morphology and physiology are to be taken into account and it is reasonable also because one factor can not be sufficient to classify about 30,000 species of Gastropoda.

13

Pearl and Pearl Industry

A pearl is a concretion produced by a mollusc. Its culture is paying industry these days. The oldest record of using pearls go back to the Chinese people. They are said to have used them as far back as 2300 B.C. Royal families and Nobles of every country in ages have adorned their persons with pearls.

What is a pearl ?

A pearl is a concretion formed by mollusc. It consists of nacre of mother of pearl as the molluscan shell. It is secreted by the mantle, the inner muscular lining of the shell, as a means of protection against foreign body. A faultless pearl is formed by the deposition of nacreous substances arranged one after the other in a concentric layer and laid down by the epithelium of mantle.

Pearl formation

The pearl formation is an adaptation against outside materials. When a foreign material such as a sand-grain or a parasite happens to enter the body it adheres with the mantle. The mantle epithelium immediately grows over the material in the form of a sac and enclose it. The mantle epithelium starts secreting concentric layers of nacre around the foreign material. The complete structure is called *pearl.*

Composition of Pearl

The pearl is formed by nacre. The nacre is formed of two substances, the calcium carbonate which is in the form of argonite or calcite and albuminoid substance called *conchiolin* ($C_{30}H_{42}N_20_{11}$). The chemical analysis of the typical pearl is :

Calcium carbonate	88-90%
Organic matter	3.8 - 5.9%
Water	2 - 3%
Residue	0.1 - 0.8%

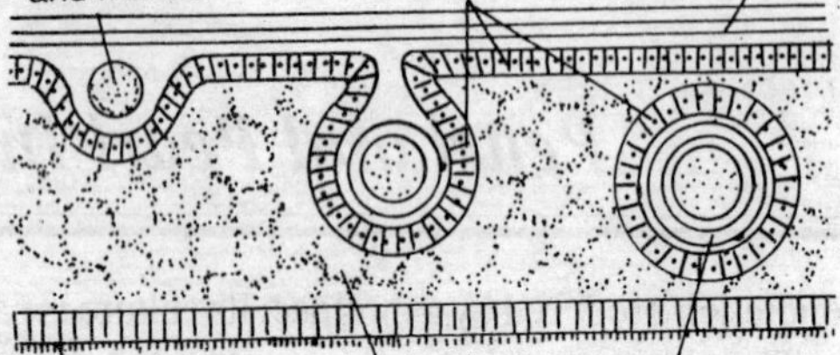

Fig. 13.1. Pearl formation.

Types of Pearls

Followings are the four main types of pearls:

1. *Blister pearls.* When some foreign matter such as sand, worms, crabs, fish or pieces of shell becomes lodged between the fleshy mantle and the shell blister pearls are formed, such pearls are attached to shell.

2. *Baroque pearls.* These pearls are most likely to be perfectly spherical, for they are formed inside the flesh of mantle where succeeding layers of nacre may be added on all sides.

3. *Hem pearls.* These are formed near the margin of mantle where the dark pigments usually produce brown or blackish pearl.

4. *Seed pearls.* The small pearls are called *seed pearls.*

5. *Cultured pearls.* These are the pearls obtained from cultured species of pearls oyesters.

Pearl Producing Animals

Pearls are produced by bivalve molluscs. These are marine as well as fresh water :

1. The fresh water mollusc is :

Unio margaritifera.

2 The marine molluscs are:

(a) *Pinctada vulgaris* (Schumacher)

(b) *P. fucata*

(c) *P. chemnitzi* (Philippi)

(d) *P. margaritifera* (Linn.)

(e) *P. anomioides* (Reeve)

(f) *P. atropurpurea* (Dunker)

(g) *Haliotis* Linn.

(h) *Mytilus* Linn.

(i) *Placuna blacenta* Linn.

(j) *P. maxima.*

The species of mollusc which is mainly is *Pinctada vulgaris* (Schumacher).

Culture of pearls

The culture of pearls is a complex but sensitive process. It involves the following steps:

1. Collection of Oysters

(i) *Japanese Method of Oyster Culture*. The larvae of the oyster freely swim the shallow waters of the sea. They are called *spats*. They are maintained in nurseries upto a certain time. For collection of spats different methods are employed. They are

(a) Semi-cylindrical tiles

(b) Bamboo poles with branches

(c) Shells of various mollucs

(d) Plastic nests

(ii) The oysters are collected from the bottom of the sea.

(iii) In the laboratories the eggs of pearl oysters are fertilized and young ones are obtained.

2. Preparation of graft tissue

The piece of tissue which is inserted into the oyster is called *graft-tissue*. The graft must be in the form of a square of 2 X 2 mm. in size and usually it is cut off form the mantle of another oyster.

3. Preparation of nucleus

The nucleus is a foreign material which is inserted into the oyster and is in the form of a bead 2 mm. in diameter. Any small particle may function as nucleus to initiate the pearl formation but it is reported that the calcarious nucleus is the best because the deposition of nacre was found to be more satisfactory on calcareous nucleus as compared to any other particle.

4. Implantation

The oyster's foot is exposed and a small incision is made on the foot. On this incision the graft tissue in placed. The nucleus is placed on the graft tissue. Then the oyster is released in the cage.

5. Rearing of oysters

The operated oysters are placed in cages and are suspended-in the sea from rafts. Rafts are rectangular wooden frames floating at the water surface. They are made to float by empty diesel drums. The rafts are kept in place by anchors. Besides raft, other methods are there, these are long-line method, rack culture, pole culture and bottom-culture method.

6. Harvesting

Pearls attain their maximum size in three years and after that the oysters are removed from cages and the pearl is taken out.

7. Cleaning of Pearls

After taking out the pearls from the Oyster shell they are washed properly, cleaned with soap solution. It is advised that pearls should not be rubbed much.

Problems of Pearl Industry

Pearl culture appears to be very much easy but a number of problems crop up during culture. Several enemies such as eels, octopus, boring sponge, star fishes etc. destory the oysters. The cold water of sea, low salinity of water, turbidity of water and high range of temperature variations hamper the pearl industry by affecting the proper rearing of the oyster.

14

Oyster culture

Oyster is an edible bivalve mollusc. Oysters are easily cultivated because of their high growth rate, sedentary habit and easy harvasting. The oysters are cultivated because their flesh is highly nutritious and palatable. The flesh contains high content of proteins and minerals like calcium, phosphorus, iron etc. Moreover, vitamin A, riboflavin, thiamine and necotinc acid are also present in sufficient amount.

The most cultivated species for this purpose are *Ostrea madrasensis, Cassostrea. gryphoides, C. discoides* and *Saccostrea cucullata.* The Indian edible oyster (*Ostrea virginiana*) is a chief food of poorer persons living on the Malabar Coast and in Cochin.

Biology of oysters

Oysters are sedentary bivalves which are found in all the seas except the colder ones. The shell is unequivalve. The left valve is large, thick and permanently attached to the rocks, piling or another shell. The right valve is smaller, thinner and flattened forming a sort of lid. The surface of the shell is coarse, irregular and ruffled. The foot is totally absent in adult but present in the larva. The left and right mantle lobes are free from each other and do not form siphons. They feed upon diatoms, organic detritus and other minute organisms. The sexes of oysters are changeable. Most young oysters are male, but half the individuals later on turn into females. Some individuals change their sex every year. An individual deposits about half a billion eggs in a season. The mortality rate is very high. The egg develops inside gill into small ciliated larva the *veliger.* These swim about and then settle at the bottom and are called the *spat.* The season of spat formation is called *spat fall* period.

Oyster culture requires selection and location of suitable site for spat collection; finding the spat-fall season; suitable methods for spat collection; growing of oysters to marketable size; clearing and marketing.

Spat collection

For the oyster culture, the spats form the seeds. From natural habitat the spats can be collected. For spat collection different materials are used. In Tamilnadu semi-cylindrical tiles are used for spat collection. The tiles are dipped in a mixture of sand and slacked lime, after that the tiles are set in water. They are laid in stacks just 50 cm. below the water surface. This method is applied before the spawning period of oysters. The spats settle down in large numbers on the tiles. Now the spats are allowed to grow upto 30 mm. in size. The spats are scraped from the tiles carefully and are transferred to growing areas. The tiles can be used for several times after cleaning and recoating with lime. Shell of various marine molluscs are also used for spat collection.

Oyster culture methods

Oysters are cultured by following methods :

(i) Raft culture

Rafts are rectangular wooden frames floating at the water surface. They are made to float by empty wooden barrels. The rafts are the anchors keep the raft floating in position. The spats are made into strings on thin ropes. The strings of spats are suspended in water from the frames. During suspension of strings, a precaution is taken that the strings not to touch the bottom.

(ii) Long-chain method

This method is the modification of raft-method. It consists of a chain of barrels under which long rope-lines are tied in a parallel fashion. The

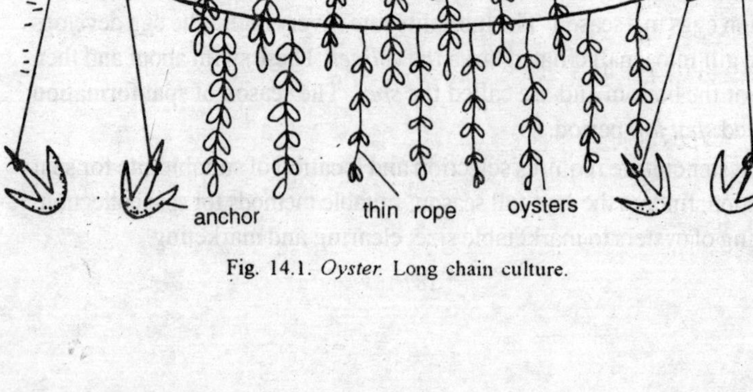

Fig. 14.1. *Oyster.* Long chain culture.

strings of spats are suspended from the ropes and the barrels are held in position by iron anchors.

(iii) Rock-culture

This method of oyster culture is employed in Tuticorin. Here two rows of vertical poles are planted from the bottom. Horizontal poles are filled to make racks. The spats are cultured on these racks by placing them or trays.

(iv) Pole-culture

A pole is planted on the bottom of sea. At the apex of the pole, a nail of 5 cm. is placed and from the nail strings of spats are suspended in water.

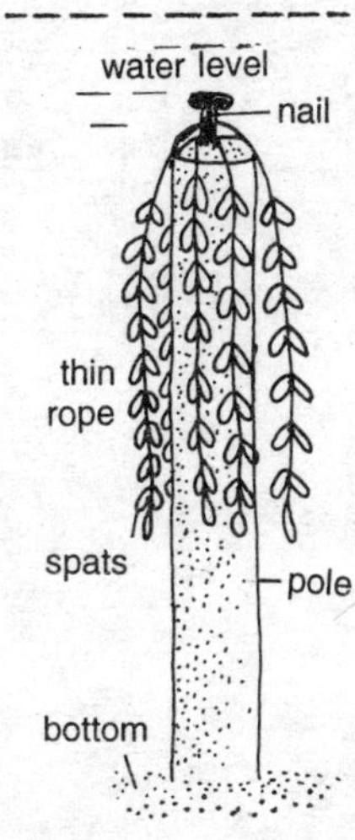

Fig. 14.2. *Oyster.* Pole culture.

(v) Bottom culture method

The empty shells of oysters are spread on the bottom and the spats are allowed to grow on them.

Harvasting

The oysters grow an average rate of 8-12 mm. per month and reach a full length of 110 mm. within one year. The oysters are harvested at this period i.e. before spawning. After spawning their flash becomes thin.

Cleaning

After harvest, the oysters are throughly cleaned with the jet of sea water. After cleaning, the oysters are placed in cemented tank filled with filtered sea-water for about one day then after they are placed in chlorinated sea water.

Marketing

After cleaning the oysters are taken to market for local consumption with shell in alive condition or by keeping in ice layers.

15

Larval Forms in Mollusca

In molluscs the development may be *direct* or *indirect*. In the direct development there being no larval stage. Here the young ones hatch form the eggs and resemble their parents in general appearance, execpt the size. The direct development is seen in cephalopods. In the indirect development a larval stage is seen during the development. Such a development is seen in majority of molluscan classes.

Paludina is a viviparous form. In some cases like *Littorina*, *Cenia* etc. the metamorphosis takes place inside the egg-case. However, three main larval forms are met with in Mollusca. These are :

1. Trochophore
2. Veliger
3. Glochidium.

1. Trochophore

The ciliated trochophore in its typical state appears in *Patella* and *Trochus* and in these forms it is formed at a very early period, even before the formation of the mesoderm, and becomes free at once. It is pear-shaped and measures about 0.5 mm. in length. In some cases they are free where as in others they are enclosed in eggs. A circlet of preoral cilia, the *prototroch* or *velum* divides the body into two unequal parts, the upper one consists of *prostomium* where as the lower part bearing *mouth* and *anus*. The preoral part is large and convex with an *apical plate* bearing the long cilia called the *apical cilia* as seen in the larvae of *Patella*, *Dentalium* etc. Near the apical cilia there are two ciliated elevations each consisting of a single cell. At the posterior end there are large anal cells bearing a bunch of cilia called *telotroch*. Between the ciliary girdle and anus, on the dorsal side, the area of ectoderm constitute the *shell glands* which secretes a thin, saucer-shaped larval shell. The trochophores of Lamellidae consists of an additional shell inside the first one which is cast off later. A slight ridge in the neighbourhood of shell represents the border of mantle.

Alimentary canal is made up of single layered epithelial cells in the form of a tube. It comprises mouth, stomodaeum, oesophagus, stomach and intestine (mesenteron). A single large cell gives rise to the mesoderm. A rudiment of radula is present in oesophagus.

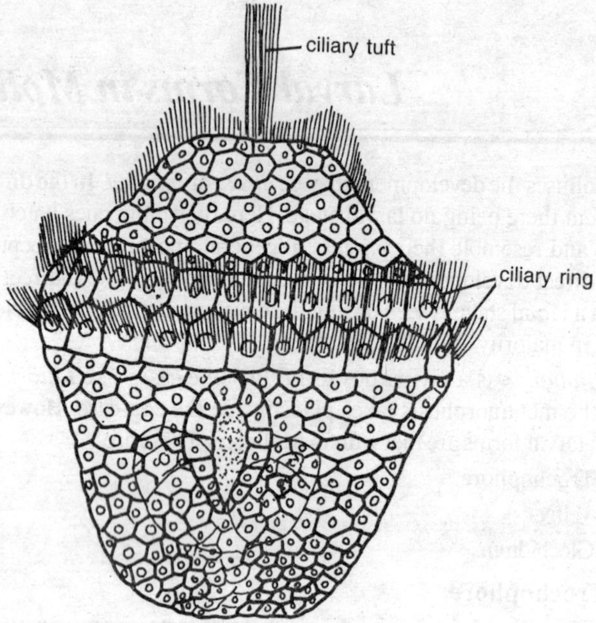

Fig. 15.1. Trochophore of *Patella*.

The trochophore contains two *cerebral ganglia*, a pair of *eyes* from the underside of apical plate.

The *statolith sacs* appear as depression of ectoderm at the sides of the mouth. Soon the trochophore develops into a yet more efficient locomotor form, the *veliger.*

2. Veliger larva

The *preoral ciliated area* or *velum* begins to protrude on both sides as a bilobed flap, that is carried in front by the larva. The velar lobes are very delicate and extensive, and a very delicate organs of locomotion. Between the bases of velar lobes the anterior end of the larva is provided with *eyes* and *tentacles.* The larva has a shell into which the velar apparatus can be withdrawn and which encloses the partly developed viscera. Alimentary canal is complete and the anus is shifted to anterior side. A foot usually bearing an operculum is present. In addition to velar apparatus, there are some organs like laıval heart and kidneys also present. The larval kidneys are paired and generally symmetrical organs situated at the anterior end of the body immediately behind the velum. Statocyst and gill-rudiments are also present.

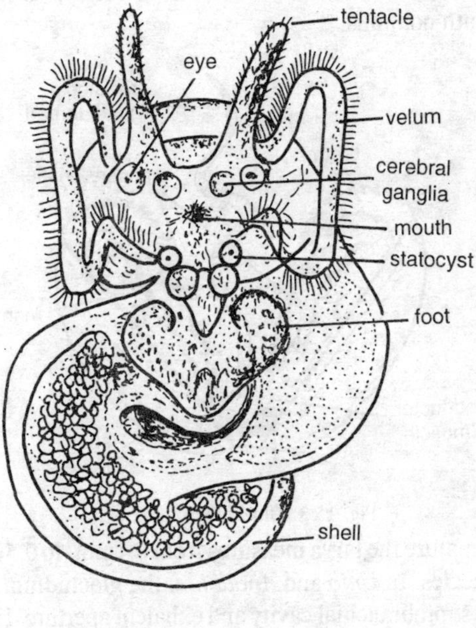

Fig. 15.2. Veliger larva of *Vermetus.*

Some gastropods have feeding veligers (planktotrophic) with a larval life that may last as long as three months. Others have a brief yolkladen nonfeeding veliger (lacithotrophic) life. The long cilia of the velum function not only in locomotion but also in suspension feeding. During the course of the veliger stage, torsion occurs and the shell and visceral mass twist 180° in relation to the head and foot.

In most of the pulmonates, the youngs are hatched out with the characters of the adult. In the species, in which veliger stage is passed in the egg, the velar lobes are reduced and the larva rotates in the egg with the help of velar cilia.

With the exception of *Dreissena* and *nausiforia*, which have a free-swimming veliger larva, fresh water bivalves exhibit modified development.

3. Glochidium larva

In the worldwide Unionacea, the Africa Mutilidae, and the South American Mycetopodidae the veliger, called as *glochidium, haustorium* and *lasidium* respectively. has become highly modified for a parasitic existence on fish. The glochidium larva is enclosed by two valves, each edge of which bears a hook. The shell valves cover a larval mantle, which bears four groups of sensory bristles. A rudimentary foot is present, to which is attached a long adhesive thread, the *byssal thread.* There is neither mouth nor anus.

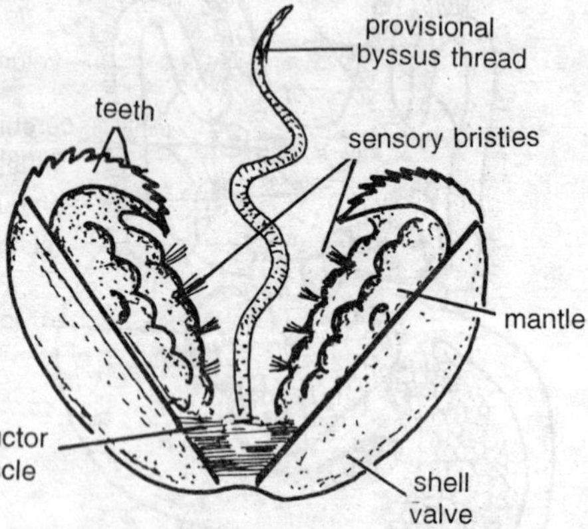

Fig. 15.3. Glochidium larva of *Unio.*

When mature the larva measures from 0.1mm. to 0.5mm. depending upon the species. In *Unio* and *Anodonta,* the glochidium leaves the gills through the suprabranchial cavity and exhalent aperture. In *Lampsilis,* the glochidia emerge directly from gills through temporary openings. The glochidia of *Anodonta* immediately clamp into the fins and other parts of body surface of the fish. The glochidia require certain species of fish as host; others can tolerate a wide range of host species and the byssal thread aids in initial contact and adhesion (*Wood,* 1974). The larval mantle contains phagocytic cells that feed on the tissue of host and obtain nutrition for the developing clam. The parasitic period lasts for 10-30 days. In the mean time the parasite is surrounded by the over growth of skin of fish forming a cyst.

Some of the larger freshwater mussels may produce as many as 3,000,000 glochidia.

16

Origin of Mollusca

The position of Phylum Mollusca is in several respects is considerable and positive isolation and any attempt to derive it from one group or other in the view of major evidences to lay any hypothesis for the origin of mollusca is ill founded. It should be noted with interest that even in such remote time as Cambrian Period, the *Sponges, Echinoderms, Platyhelminths* and *Mollusc* have already been identified as separate and distinct groups with the same distinction from each other what we observe in the present time. In such circumstances any system of classification or any theory of development which proposes the derivation of Mollusca from any of this group is extremely dangerous.

It would, therefore, almost certain that the origin of Mollusca is merely a metaspeculation and if at all any attempt is to be made we have to examine the indication of the relationship of the Phylum by the consideration of embryonic resemblance. One of the most important feature of Phylum Mollusca is the occurrence of a *trochophore larva* in the development. It is interesting to note that such a larval form is also shared by annelids but if the presence of trochophore is in any form to be regarded as important in shading some light from the origin of Mollusca, it should not necessarily be taken as indicative of an annelid ancestor. If we examine the trochophore larva of any mollusc we see a complete gastrulation and the ciliated bands besides a single large cell, which would give rise to the mesoderm in the successive developmental stages is seen. In addition to the position of an apical organ i.e. *protoroch* and larval nephridia its internal organisation also shows similarities with the annelidan trochophore. Such apparent similarities appeal to be discussed and a more careful study would be certain striking differences between the two which can not escape our attention. We thus say on the dorsal side of advanced molluscan trochophore there is a rudiment of mantle formed by ectoderm. At this stage the earliest sign of the shell can also be seen.

The food has also made its appearence as a ventral prominence. The single large mesodermal cell gives rise to two bands of mesodermal cells in each of which a cavity develops, but by for the most important difference from annelid is afforded by the absence of segmentation in the embryonic mesoderm of mollusc. It would thus appear that absence of segmentation provides a positive evidence from such group of animals which were devoid of segmentation.

From the foregoing account if becomes positively certain that we have to leave our such view for the origin of Mollusca from some annelid type ancestor may help us. Such a connection is proved by the striking similarities that are present between the ladder-like nervous system in certain turbellarians and in placophore, solenogasters and certain aspidobranch gastropods. In either case we find that the ganglia itself are distributed along the entire length of the main nerve trunk. The pleurovisceral and pedal nerve cords should be taken to correspond to the lateral and ventral nerve cords of the turbellarians. Now if such a hypothetical ancestor could suppose to has given rise a primitive mollusc then we should have to postulate following important changes that should have undergone in its organisation for such a purpose:

1. Secretion of dorsal shell, which in the begining might have been in the form of a layer of cuticle, containing calcareous particles and would have served as a protective covering.

2. The development of shell would evidently reduced the general respiratory surface of the body which in turn would make the necessity to develop such specialized respiratory organ as ctenidium as present in molluscs.

3. The ventral musculature which have already been well developed in the turbellarian must have undergone a remarkable development to give rise to such a specialized and efficient molluscan foot.

It is worthy of notice that typical trochophore larva bears close resemblances to certain form of turbellarian larva i.e. *Muller's larva.* If we have to assume that the primitive molluscs derived from some turbellarian like ancestor we will also have to postulate the converson of Muller's larva into a molluscan trochophore. Such a postulation would automatically lead us to the contentation that the annelids and molluscs had probably common platyhelminthic origin will subsequent extreme divergence into two phyla. If ontogeny repeats phylogeny then a trochophore will be taken to be present a remote ancestor of annelids and molluscs which resembles the ancestors of platyhelminthes,represented by Muller' larva.

The trochophore larva by certain changes developed into annelidon form. These changes include :

1. Elongation of body which becomes segmented.
2. Formation of septa.
3. Development of coelom.

On the other hand transformation into molluscan form occurred through the following changes :

1. The apical cilia got transformed into velum which serve not only for locomotion but also for feeding. The cilia of apical plate create a current of water which bring food particles into the mouth.
2. Shell gland got developed which secretes the unique shell.

But in 1952 the whole picture changed when 10 living specimens of *Neopilina* were found from a depth of 3500 metres off the Paciffic coast of Costa Rica (Maxico). The most unusual feature of *Neopilina* which belongs to the Monoplacophora of Mollusca in its appearence, metameric plane of structure. It displaces 5 pairs of gills, 6 pairs of nephridia and these organs are more or less metamerically arranged, a number of zoologists believe that this metameric plane displaced from *Neopilina* is an evidence that

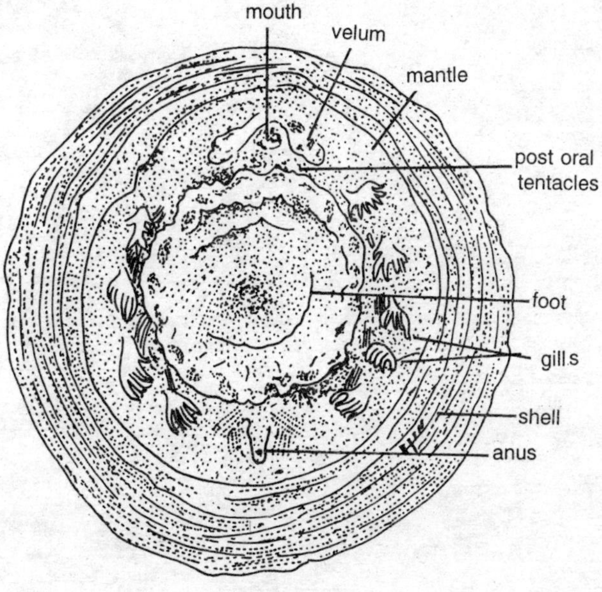

Fig. 16.1. *Neopilina galatheae.*

Mollusca comprise a fundamental metameric phylum and must have evolved from the annelids. The molluscan trochophore is thus believed to be additional evidence of annelid origin of mollusc.

In the light of *Neopilina* it may be existed to revise our thinking in future, but at present the older thing assuming a common origin of annelid and mollusc seems to possess the greatest weight of evidence.

17

Evolution of various classes of Mollusca

From the primitive archimolluscal structure, different groups known to-day, have diverged and each of the groups follows the evolution along certain lines of specialization.

1. Amphineura or Placophora

The amphineurans are the *Chitons*, seem to have departed least from the ancestral mollusc. Their body became elongated and protected by eight shell-pieces on the dorsal side. On either side of mantle cavity was symmetrically arranged several pairs of bipectinate gills. If we were to assume that the primitive mollusc possessed several ctenidia, then it is appearent that the chitons have retained the primitive characters. On the other hand if we assume that the primitive mollusc possessed a single pair of ctenidia then it naturally employes that the numerous ctenidia of Chitons are formed by the repeatitior of the former. The foot in Chiton has retained the primitive from. The nervous system is also primitive in that the ganglionic cells are distributed along the whole length of nerve cords. Chitons also possess a well developed radula.

2. Gastropoda

The leading features in the evolution of Gastropoda has been the torsion of visceral mass producing asymmetry in body. There is no doubt that the ancester gastropod possessed a body with a straight alimentary canal, leading in a posteriorly placed anus, and on either side of anus was placed the ctenidium. We also know well that gastropods have univalved shell which covers all the soft parts of body and extension of foot on the ventral side tend to remove the pallial complex with and renal openings away from head. The univalved shell, therefore, meant appearently necessary for mantle cavity, ctenidia and excretory aperture to shift at anterior and somewhat in the neighbourhood of opening of shell. The whole of pallial complex was shifted from its primitive posterior position to the anterior position by the torsion. The large majority of gastropods included in order Prosobranchiata show torsion in full development. It is also known that each group of Mollusca shows a progressive loss of 'shell in the course of evolution. It has also been invariably observed in Gastropoda that such a loss of shell is accomplished by a reversion of the phenomenon of torsion i.e. detorsion and consequent rotation of visceral loop back to its posterior condition i.e. in opistbobranachiates. The opisthobrnachiates are derived from Pectinibranchiata is prooved by the fact that they possess a single ctenidium, single auricle, besides embryological evidences also support such a view. Even those opisthobranchs which are detorted in adult condition the primitive torsion is seen in the course of development. The nudibranches seem to have more step further. Here the detorsion is complete and have lost the ctenidium as well as the shell. But they have evolved from a fully torted gastropods when we study their embryology.

We have tried to show a gradual evolution of different groups of Gastropoda on the basis of torsion and its subsequent reverse, but nervous system also shows a way to the evolutionary history of Gastropoda. We find the most primitive type of nervous system in this class which is exhibited by certain aspidobranchs is that they show a lesser tendency towards the concentration of various ganglia, besides they possess longitudinal pedal cords which can be compared to pedal cords of *Chiton*, but the evolution of nervous system in this class has taken along two lines:

1. In the establishment of progressive euthyneury are symmetry.
2. In the concentration of various ganglia and their connectives around the circum-oesophageal nerve ring. The visceral loop remains twisted and in all the prosobranchiates are the result of torsion.

The opisthobranchiates showing detorsion have an untwisted visceral loop, a concentration of various ganglia thus begin in the cephalospides for example in *Bulla* the pleural ganglia shift very close to the cerebral ganglia or they may be fused as in *Aceton*. In *Notochus* the pleural- visceral connectives are considerable shortened. In the nudibranchs the tendency of concentration of various ganglia and shortening of various connectives have gone still further. Taking for example in the case of *Tethys* we find that the pleural, pedal and cerebral ganglia of each side fuse together in the form of compect pleuro-cerebral-pedal mass and with its counter part on other side it forms a large supra-oesophageal ganglionic mass.

So far we have excluded the pulmonates for any discussion. The typical member of *Pulmonata* shows a coiled visceral mass possessing a shell and exhibits an internal streptoneural asymmetry to certain extent, but this should not lead us to believe that these are a group of less specialized forms. On the other hand, they exhibit a perfect symmetry of nervous system, the concentration of various ganglia and shortening of various connectives. The final result is, therefore, incorporation of visceral loop together with its ganglia and circum-oesophageal nerve cord.

On an examination of respiratory organs, one finds a gradual loss of ctenidia as we move from lower groups to the higher ones in class *Gastropoda*. We, thus see that certain aspidobranchs (*Haliotis*) exhibit the most primitive condition as they retain a'pair of bipectinate ctenidia. Other aspidobranchs belong to family Trochidae, Tuberidae and Neritidae have only one ctenidium and that too is bipectinate. From this bipectinate type of gills arose monopectinate type of gills of pectinibranchs which have the last leaf-lets of one side of the animals. Certain land prosobranchiates have lost their ctenidia in relation to terrestrial life and consequently respire with the help of pulmonary sac. In *Pila* while still retained the monopectinate gill has developed a pulmonary cavity and is capable of leading an amphibious life. When a true ctenidium is present among opisthobranchiates it lies posteriorly on right side as a result of detorsion. The nudibranchs have lost the true ctenidium as well as mantle cavity and respiration is carried with the help of secondary gills or outer skin. Finally in Pulmonata the ctenidia are totally absent and the respiration is effected with the help of pulmonary sac.

So for as the foot in gastropods is concerned, in Aspidobranchiata it is very well developed and is differentiated into propodium, metapodium and epipodium. All of which show remarkable development. In Pectinibranchiata epipodium is usually absent, in certain prosobranchiates the foot may be reduced in simple discoidal prominence (*Vermetus*). It may be represented by small, ventral appendage in parasitic forms. Passing to Opisthobranchiata we find that there is no division of foot of various forms. The parapodia may be modified into fold-like extensions from the edges of the sole of foot and bent back over the shell. Among certain nudibranchs the foot no longer exists as a differentiated organ while pulmonate show poor development.

From what has been said above it can be safely concluded that the most primitive gastropod, Aspidobranchiata, which in turn gave rise to the Pectinibranchiata. The Pectinibranchiata by a series of changes in their internal as well as external organisation gave rise on one end to the Opisthobranchiata and on the other Pulmonata and have made specialization among two different lines.

3. Pelecypoda

As far as Bivalvia is concerned the fact that bivalved shell at first appearance is univalve and that of the foot of Protobranchiata is of creeping type and the ctenidia are plume or feather-like suggest the derivation of their class from animal resembling a simple gastropod with no radula and undisturbed bilateral symmetry. On the other hand certain zoologists record this proof to have been derived from the hypothetical primitive ancestor with a more or less developed head and radula. Such a primitive mollusc with the process of degeneration on account of leading an inactive sedentary life was modified accordingly either lost or fail to acquire those specialized characters of the organization of highly developed forms. The pelecypods thus arose with absence of any definite region and masticatory apparatus.

From the point of view of physiology, the most primitive pelecypods are those in which the foot has a creeping ventral surface, and the branchial filaments are free and not reflected those of the protobranchiate. From these are evolved the filibranchiates where branchial filaments are reflected but are sfill lacking vascular function. These is their term gave rise to the eulamellibranchiates which are most specilized in respect of complication of ctenidia. Numerous vascular inter-lammellar and inter-filamentar junctions are present. From eulamellibranchiates are derived septibranchiates in which respiratory organ is represented by branchial septum.

The most primitive condition of nervous system exhibited by Protobranchiata as the cerebral and pleural ganglia are still separated and there is a set of cerebro-pleural, cerebro-pedal and pleuro-pedalconnectives on each side. In all other lamellibranchiates, however, the cerebro-pleural ganglia are intimately fused to one another and there is a single pedal connective on each side. The two pedal ganglia also lie close to each other and may be reduced in those forms in which the foot is reduced or absent e.g. *Pecten*, *Teredo*, *Ostrea* etc.

4. Cephalopoda

The cephalopods are bilaterally symmetrical and have a posterior mantle cavity. The primitive form in this group i.e. tetrabranchiates possess an external shell which may be many chambered besides they possess two pairs of ctenidia, kidneys and auricles in contrast to more specialized form i.e. Dibranchiata which have each of these structures in single pair and have an internal shell which may be reduced or absent. The remarkable features of cephalopods, however, is extra-ordinary modification of foot into arms and tentacles. This class remains apart from molluscs by its wonderfully high organisation, specialized nervous system and eyes. There is nothing to indicate close relationship with any of lower groups beyond the general resemblance to molluscan organisation and presence of odontophore.

The cephalopods form, infact, a singularray isolated group. Palaeontology has also not given any indication of their origin. Embryology is equally silent. The absence of free larva and performed modification in the development produced by enormous mass of food-yolk sharply separates them from all the members of Phylum Mollusca.

18

Adaptive Radiation

Because of the constant competition for food and living space, each group of organisms tends to spread out and occupy as many different habitats as possible. This process of evolution from a single ancestral species of a variety of forms that occupy somewhat different habitats is termed *adaptive radiation*. It is clerly advantageous in evolution in enabling the organism to trap new sources of food or to escape from some of their enemies. The need for food and safety is given the primary cause of adaptive radiation.

Lamellibranchs are the main bivalve molluscs

ADAPTIVE RADIATION OF BIVALVES

The evolution of filter feeding freed lamellibranchs from dependence upon deposit material and made possible the colonization of many habitats that were uninhabitable for their protobranch ancestors. The success of this adaptive radiation is reflected in the fact that of the some 20,000 described species and 75 families of bivalves, most are lamellibranchs.

It must be emphasized that the adaptive groups described below do not necessarily constitute closely related species. Colonization and adaptiation for a particular type of habitat has been achieved independently by a number of different bivalve families or superfamilies.

Soft Bottom Burrowers (Infauna)

The majority of lamellibranchs are inhabitants of soft bottoms, exploiting the protection offered by a subterranean life in marine sand and mud while utilizing food suspended in water brought in from above the surface. Some live just beneath the surface; some are adapted for deep burrowing; some move between the surface and lower levels; and some are especially adapted for shallow, rapid burrowing in a shifting environment. In addition to the many marine forms, soft bottom lamellibranchs also include most of the freshwater clams and mussels.

Mechanism of burrowing

The actual mechanism of burrowing involves the coordinated action of a number of forces. The protraction of the foot is initiated by contraction of the pair of pedal protractor muscles. These muscles extend from each side of the foot to the adjacent valve, to become attached just beneath the anterior adductor muscle. The projecting foot probes and pushes into the surrounding sand. As the foot is protruded, the valves begin closing by contraction of the adductor muscles. Some water from the mantle cavity is expelled, which loosens the sand or mud and facilitates the movement of the foot. The water remaining in the mantle cavity and the blood act as a hydrostatic skeleton. The pressure of these two fluids is elevated by the adducted valves. Blood from the visceral mass is forced down into the pedal hemocoel, causing the foot to dilate and anchor into the substratum. With the foot anchored, an anterior pair and a posterior pair of pedal retractor muscles contract, pulling the shell downward. The pedal retractors also originate on the shell in the region of the adductor muscles. In many species, retraction by the posterior muscle. The effect is to rock the shell that facilitates its movement through the substratum. Ridges or other sculpturing of the shell surface may increase traction. Following pedal retraction, relaxation occurs and the valves gape. To return to the surface or to a higher level within the substratum, some soft bottom bivalves actually turn around and burrow upward, but most back out, pushing again the anchored end of the foot.

(a) Burrowing in Protobranchs

In the primitive protobranchs, such as *Nucula, Solemya,* and *Yoldia,* the foot bears a little flattened sole. However, the two sides of the sole can be folded together, producing a bladelike edge; in this condition the foot is thrust into the mud or sand. The sole then opens and serves as an anchor and the remainder of the body is pulled downward after the sole. *Solemya,* a rapid mud burrower, can completely disappear under the mud surface with two thrusts of the foot.

(b) Burrwoing in common cockles

The common cockles (*Cardium* and allies), which have rounded, convex ribbed shells, are shallow burrowers. They use the foot not only for conventional burrowing but also for escape. When threatened by a predator, such as a starfish, the cockle leaps by rapidly extending the foot from a folded position against the substratum.

Problems in burrowing

A major problem arising from burrowing in soft bottoms is that of sediment brought in by water currents. Blood engorgement within the

mantle margin enables the mantle edges to be appressed together, even when the valves gape slightly (the muscles of the inner lobe provide for retraction, but since circulation of water through the mantle cavity is necessary for gas exchange and, in most species, for feeding, at least the posterior part of the mantle must be opened for entrance of water, with which sediment enters.

There has been a tendency to seal the mantle edges morphologically where openings are not necessary. The most frequent point of fusion of the mantle edges is at the posterior between the inhalant and exhalant openings. This fusion forms a distinct exhalent aperture. A functional inhalant opening is present since the mantle edges are pressed together below the opening, although the mantle edges are not morphologically fused. This type of structure is found in the protobranch *Nuculana* and in the freshwater lamellibranch clams *Anodonta* and *Unio.* A second point of fusion just below the inhalant opening, forming a distinct inhalant aperature, has evolved in many bivalves.

Still further mantle fusion has developed in some species, especially deep burrowers, and most of the ventral margin anterior to the inhalant aperture has become sealed. Thus, three apertures remain—the inhalant and exhalant apertures and an anterior pedal aperture through which the foot protrudes. *Mya* and the razor clams *Ensis* and *Tagelus* are examples of bivalves with pedal apertures. Extensive mantle fasion not only contributes to the reduction of fouling but is also important in maintaining the hydraulic pressure within the mantle cavity necessary for burrowing.

Commonly, when structural inhalant and exhalant apertures are present as a result of fusion, the mantle edges surrounding the apertures have become elongated to form tubular siphons of varying lengths. With siphons the animals can be completely buried in the mud and only the siphons are extended by blood pressure or by water pressure within the mantle cavity when the valves are closed and are withdrawn by siphon retractor muscles derived from the muscle tissue of the innermost mantle fold. Considerable variation exists in siphon length and degree of fusion between inhalant and exhalant siphons. Where there are separate inhalant and exhalant siphons, siphon formation involves only the muscular lobe of the mantle. Fused siphons may involve the sensory or shell-secreting lobe as well.

When well-developed siphons are present, they are indicated by scars on the inner face of the valve. The pallial line impression is recured sharply inward just below the posterior adductor and represents the point of attachment of the siphon retractor muscles. This bay in the pallial line is called the *pallial sinus.*

The radiation of filter feeding bivalves with siphons did not occur until the Mesozoic and Cenozoic, but it led to the development of many groups that invaded not only deeper layers of the substratum but offshore bottoms as well, habitats from which Paleozoic bivalves were largely absent. Most modern soft bottom filter feeders possess siphons. Those that do not are rather sluggish shallow burrowers.

The geoducks of the Pacific coast of the United States, *Panopea generosa,* are among the deepest burrowers, extending down to a depth of over a meter. They have siphons so large that they can no longer be retracted.

Many bivalves that burrow deeply (i.e. to depths greater than the lengths of their own bodies) tend to have semipermanent or even permanent burrows. The walls of the burrows become coated with mucous, which also reduces sediment fouling.

The primitive form of lamellibranch soft bottom burrowers is considered to be one in which the adductor muscles are more or less equal (isomyarian), the mantle is unfused, and the valves are equal and circular in outline. In many groups the valves are also modified, becoming more streamlined for burrowing by being flattened or elongated. Species of *Donax,* which are inhabitants of surf beaches, have shells that are pointed anteriorly and blunt posteriorly. They back out on incoming waves and reburrow with great rapidity as the wave recedes. The margins of the inhalant siphon are fringed with infolded tentacles, which keep out swirling sand grains.

The razor clams *Ensis* and *Tagelus,* two unrelated but similar groups, have greatly elongated valves and an elongate foot, which enable them to move rapidly within their more or less permanent burrows. *Tagelus* has long siphons, each of which has a separate opening to the surface. *Ensis,* which has short siphons, comes to the surface to feed from the deeper, more protected part of its burrow.

Two further specializations of soft bottom burrowers should be mentioned. Some tellinaceans have reverted to deposit feeding. *Scrobicularia,* for example, extends its inhalant siphon above the surface at low tide and, like a vacuum cleaner, sucks in deposit material, which is then sorted on the gills. Deposit feeding is generally utilized as an addition to, rather than as a substitute for, filter feeding.

The lucines, rather small rounded shells, (*Linga, Codakia, Lucina, Divaricella, Thyasina*), are remarkable in using the foot not only for burrowing but also for constructing a mucous-lined canal back up to the surface. The feeding ventilating current enters anteriorly by this passageway and leaves by the usual exhalant opening.

Molluscs attached with form substratum or attached Surface Dwellers

A number of evolutionary lines of lamellibranchs have invaded form substrates—peat, wood, shell, coral, rock, or man-made sea walls, jetties, and pilings. Attachment is provided either by byssal threads or by one valve fused to the substratum. Byssal threads are tough, horny threads secreted by a gland in the foot. The secretion flows out to the substratum along a groove in the foot. Then after the fiber hardens by tanning, the foot is removed, leaving the thread anchored to the substratum at one end and to the byssus opening of the foot at the other. A byssus retractor muscle may enable the animal to pull against its anchorage.

Among living surface dwellers attached by byssal thread, the widely distributed mussels, members of the family Mytilidae, are perhaps the most familiar. They live attached to wharf pilings, sea walls, and rocks or among oysters, often in great numbers. The threads, laid down by the little finger-like foot, often radiate outward like guy wire. Young individuals are even able to use byssal threads to climb vertical walls. Mytilids are widely used as food in many parts of the world and are sometimes "farmed" on ropes suspended from rafts.

Other surface inhabitants attached by byssal threads include many of the heavy-bodied ark shells which are very common on tropical coralline substrates; mangrove oysters (*Isognomon*), which hang in clusters from mangroove roots; and winged oysters (pteriids), which live attached to sea fans and other gorgonian corals.

The Tidacnidae, which include the giant clams (*Tridacna*) of the Indo-Pacific, are also surface dwellers. The smallest of the six species is only 10 cm. long; the largest *T.gigas*, reaches 1.37 meters. All live vertically oriented with the hinge side down and are initially attached by byssal threads. Some retain the byssus; some lose it and rest on the bottom by the weight of the shell; *T.maxima* bores in coral or coralline rock so that the valve margins are flush with the substrate surface. The gape of all tridacnids is directed upward and with the large mantle surface (actually the siphons) protruded across the fluted shell for maximum exposure to light. The mantle tissue contains symbiotic zooxanthellae that provide the clam with an accessory source of nutrition. At night zooxanthellae are removed from the mantle by amaebocytes, in which they are digested. Undigestible remains are then passed to the digestive glands and nephridia for removal. The mantle tissue also contains pigments of brilliant green, blue red, violet, or brown hues that probably function to reduce light intensity. Large populations of tridacnids, such as the little boring *T. maxima*, can be a spectacular addition to the beauty of an Indo-Pacific coral reef.

Utilization of byssal threads by adults of some 18 different bivalve superfamilies is believed to represent a persistent larval adaptation, for the larvae of many unattached burrowing forms produce a byssus for initial temporary anchorage on settling. Adults of a few living species use byssal threads to anchor rootlike in soft bottoms. The mytilid genus *Modiolus* live partially buried in peat or coarse substrata, and the hummer oyster *Malleus* and the pen shells *Pinna* and *Atrina* occupy a similar position in sand, attaching the byssal threads to small stones. However, such an endobyssate habit was a common adaptation of Paleozoic bivalves (40 per cent of Ordovician bivalves were endobyssate), and *Stanley* (1972) believes that adult retention of byssal threads for stabilization in soft sbustata was the evolutionary rounte that lead to the more common bysally attached surface habit (epibyssate) encountered today. Although most burrowing forms evolved from shallow burrowing ancestors described earlier, *Stanley* believes that some burrowing groups may have returned to a burrowing habit from endobyssate ancestors.

Surface dwelling bivalves that are attached by cementation lie on one isde, fixed to the substratum by either the right or the left valve, depending on the species. Such sessile bivalves include four different evolutionary lines, of which the oysters are the most familiar. However, the name oyster is applied to a wide variety of speices, some of which attach by byssal threads. In the family Ostreidae, which contains the edible American east coast oyster *Grassostrea virginica* and the European *Ostrea edulis,* the metamporphosing veliger is initially anchored with a drop of adhesive fluid from the byssus gland upon settling. Then the mantle margin attaches the left valve to the substratum in the process of shell secretion. Shell attachment has led to varying degrees of inequality in the size of the two valves, the lower being larger or smaller than the upper one. In *Chama,* the topical jewel boxes, the upper valve forms a lid over the boxlike lower valve. The extreme condition was reached in the extinct Mesozoic rudists, in which the lower valve was shaped like tube or horn. The rudists often occured in reeflike aggregations.

The common jingle shells or toenail shells, members of the family Anomiidae, possess features of both the byssally attached and the cemented bivalves. They lie on one side but are actually anchored by a large calcified byssal thread that passes through the attached valve.

Attached bivalves share a number of features. As would be expected, the foot is reduced to varying degrees, although not in the primitive ark shells, and it is completely absent in those bivalves, such as the oysters, that are attached by one valve. There has also been a tendency for the anterior end to become smaller, leading to a reduction (anisomyarian) or loss (monomyarian) of the anterior adductor muscle. It has been suggested that anterior reduction in mussels (Mytilidae) is perhaps an adaptation to elevate the posterior end, thereby reducing the likelihood of obstruction that the dense aggregations of these bivalves might create. Not unexpectedly, many sessile bivalves, such as various scallops and thorny oysters, have the sensory lobe well developed with tentacles and eyes.

Mantle fusion and siphon formation have not occured in sessile bivalves, since they are above the surface and generally on hard substrata, where sedimentation is less of a problem. However, oysters and mussels, which occur in dense beds, are dependent upon the cleansing action of tidal currents to prevent the animals from becoming completely buried in their own faeces and pseudofaeces. Young *Mytilis edulis* actually clean the exterior of their shells with their long foot.

MOLLUSCS LIVING ON THE SURFACE

Unattached surface Dwellers

Among the small number of bivalves that live free on the surface, some scallops (Pectinidae) and some file shells (Limidae) are the most familiar. There are members of both families that live anchored by byssal threads, but others are unattached, or only weakly attached. They lie on their sides, and in many scallops that valve (right) is flattened; the foot is reduced, and the anterior adductor muscle has disappeared. Free-living file shells and scallops have evolved the ability to swim by rapid ejection of water from the mantle cavity with the clapping of the valves. The solitary posterior adductor muscle has shifted to a more central position and is divided into smooth and striated sections. The rapid contraction of the striated fibers provides for swimming, and the sustained contraction of the smooth fibers provides for prolonged clousre of the valves. The muscular lobe of the mantle margin, when appressed against the lobe on the opposite mantle surface, controls the direction of the water jet, permitting it (in scallops) to exit on either side of the hinge line or opposite the hinge line. The mantle sensory lobe is also highly developed, bearing tentacles and, in scallops, many small blue eyes.

The swimming ability of scallops and file shells is used primarily to escape predators or other sudden distrubing conditions. For example, if a predatory starfish, or even one tubefoot of such a starfish, touches the mantle margins of a scallop, it will evoke a swimming response and the scallop will be propelled a meter or so away. Some scallops use the water jets to blow out a depression in the sand surface into which they settle.

File shells typically nest in crevices beneath stones and only swim when disturbed.

Other free surface dwellers include some odd jingle shells, *Enigmonia enigmatica* (Anomiidae), which live on mangrove trees in Australia. Unlike other jingle shells, which are highly adapted for an attached existence, these species can crawl up and down the mangrove trunks throughout life by means of a gastropod-like foot, which protrudes through a hole in the shell, but temporally attach by means of a byssus for a long as 10 days. A related genus, *Palcuma,* contains species, called window pane shells, which live free on the surface of mud flats in the Indian Ocean.

Boring Bivalves

The ability to penetrate and live beneath the surface of firm substrate—peat, clay, sandstone, shell, coral, coralline rock, and wood—has evolved in seven super-families of lamellibranchs. The ancestors of some of these boring bivalves were soft bottom inhabitants that evolved the ability to burrow in successively firmer substrates. the ancestors of others were surface inhabitants attached to hard substrates by byssal threads.

All boring bivalves begin excavation following settling of the larva and slowly enlarge and deepen the burrow with growth. They are forever locked within their burrows, and only the siphons project to the small surface opening. If a boring bivalve is removed and placed on the surface, it cannot excavate a new chamber.

Boring bivalves with epifaunal ancestors hold to the side of the burrow by byssal threads. Boring species with burrowing ancestors usually attach by the ventral surface of the foot, which has developed a sucker-like surface. In the great majority of species, drilling is a mechanical process and the anterior ends of the valves, which are frequently serrated, are the abrading surfaces. The drilling movements are commonly adaptations of the movements found in their nonboring ancestors. Depending upon the sepcies, cutting force results from : (1) an upward movement of the anterior end of valves due to contraction of posterior pedal retractors (*Petricola, Hiatella*); (2) a downward movement of the anterior end of valves due to contraction of anterior pedal retractors (*Gastrochaena*); (3) an opening thrust of the anterior ends of the valves due to contraction of the posterior adductor muscle (*Platyodon,* pholads); (4) a back and forth movement due to alternate contraction of the anterior and posterior pedal contractors pulling against two groups of byssus (*Botula*). Modifications to increase valve articulation and cutting movements include reduction of the hinge ligament and a shift of the anterior adductor muscle to a position *above* the hinge line. This region may then by covered by accessory shell plates.

Some boring bivalves rotate within the burrow, i.e., change position, and as a result the burrow cross-section is round. Others remain attached in one place, and the burrow tube takes the shape of the shell. Such forms may even have two openings to the surface if the ends of the siphons secrete some material between them (*Gastrochaena*). The sediment produced in drilling is eliminated by taking it into the mantle cavity and then ejecting it with the pseudofeces through the gut.

Lithophaga, a very common cigar-shaped byssally attached borer in shell and coral, excavates chemically. An acid-mucous secreted by the mantle margin softens the calcareous substratum, which is then scraped away with the valves. The shell of *Lithophaga* is protected from its own mucous by the periostracum.

Many boring bivalves inhabit wood. The wood-boring Pholadidae, such as *Martesia* and *Xylophaga,* are adapted in much the same way as the many rock and shell-boring members of the same family. Wood panels planted on the deep sea bottom at 1830 meters south of Woods Hole, Massachusetts, by the submersible research vessel *Alvin* were completely riddled by *Xylophaga* and a related genus when recovered 104 days later. The most specialized wood borers are the shipworms, members of the family Teredinidae. The natural habitats of the some 60 species of this widely distributed family are mangrove roots and timber swept into the sea by rivers, and they play an important ecological role in the reduction of sea-borne wood. They are a serious pest of piers, pilings, and other wooden structures placed by man in the sea, and much expense and research has been devoted to antifouling measures.

The body of the shipworm (*Teredo*) is greatly elongated and cylindrical. The shell is reduced to two small anterior valves. Cutting of the wood is effected by opening and rocking motions of the valves while the anterior end of the body behind the valves, produces a calcareous lining within the tunnel. The long, delicate siphons open at the surface of the wood, and the burrow entrance is plugged by calcareous pallets when the siphons are retracted. The burrow increases with the growth of the shipworm that fills it, and may reach a length of 18 cm. to 2 m., depending on the species. The life span is one to several years, again depending on the speices. Timbers can become completely riddled with tunnels.

Shipworms use the excavated sawdust for food. The stomach is provided with a caecum for sawdust storage, and a section of the digestive gland is specialized for handling wood particles. Cellulose digestion probably occurs intracellularly within the digestive gland and in some species extracellular by symbiotic bacteria housed within a secial organ that opens into the oesophagus. In these species for which sawdust is the principal food, nitrogen-fixing gut bacteria compensate for the low protein levels in the diet. The relative importance of sawdust and plankton as food is reflected in the size of the caecum and the gills in different groups of shipworms.

Commensals and Parasites

A small number of bivalves have evolved a commensal or parasitic relationship. The commensals are almost all members of the superfamily Erycinacea, which also contains free-living epibenthic forms, such as the little *Kellia* and *Lasaea* that nestle in crevices. Most attach by byssus threads, but some crawl on the foot like a snail. The hosts are usually burrowing echinoderms such as heart urchins, brittle stars, sea cucumbers, and burrowing shrimp-like crustaceans. *Entovalva,* which lives in the gut of sea cucumbers, is the only known parasitic bivalve.

19

Economic Importance of Mollusca

Molluscs are useful as well harmful. They have been a great economic importance since very long. They have been used by for various purposes.

USEFUL MOLLUSCS

As food

As food and bait many molluscs are very useful to man. They served as food for ancient man and primitive peoples. The large land snail, *Helix pomatia* is eaten in Southern Europe. The French, Italians and Spanish eat *Natica, Turbo, Triton* and *Murex*. The apple snail, *Pila globosa* is regularly eaten by the poor people in the Tanjavur and Tirunelveli districts of South India.

Bivalves furnish some of the most delicious and nutritious foods man has discovered. Oysters, *Venus, Lithodomus, Pholas* etc. are eaten by Franch and Italian people. The Americans eat *Venus, Mya, Maltra.* American oyster, *Crassostrea viginica* is commercially cultured and harvested in a very efficient manner with millions of dollors are involved in this industry. The adductor muscles of *Pecten* are served rolled in flour and fried. The malayans of East Indian islands eat *Haliotis* and *Telescopium.* The lower classes in Phillipines eat *Arca.* In Corean Islands, *Monodonta* and *Mytilus* are eaten. West Indies and Negroes eat the large chitons and the chitons are commonly known as *sea-beef. Anodonta edulis* is specially cultivated for food in some parts of China.

Cuttle-fish (*Sepia*), squids (*Loligo*) and devel-fish (*Octopus*) are popular articles of the human diet in Oriental and Mediterranean countries. In Japan, India, China and Italy, they are sold in the markets for food. Sepia is abundent in European waters. Red-Indians, dwelling along the Pacific Coast in Canada and Alaska, are said to use the *Octopus* on many occasions. *Nautilus* is much prized as food by Pacific islanders.

Pelycepods furnish food for numerous other animals. Their chief enemies are star-fishes, boring sponge, drilling snails, some marine leeches, fishes and shore-birds.

Cephalopods are also an important source of food for other animals. The giant squid (*Architeuthis*) of cold depths of New Foundland are eaten by sperm whales.

2. As a bait for fishing

In England specially in Scottland, mussels are largely used as bait for fishing. *Whelks* are largely employed for bait, specially in the cod-fishery. Squids make an excellent bait for marine fishes.

3. Shell as money

The gastropod-shells provide the money of various native races. One of the commonest species is the cowry (*Cyprea moneta*), which alone is used as entire while nearly all other forms of shell money are made out of portions of skills. A handed form of *Nerita polita* was used as money is certain parts of South Pacific. The tooth shell of *Dentalium* is commonly used as money by Red Indian tribes of America.

4. As ornaments

Shells are used to form ornaments by savage and even by civilised tribes in all the parts of world. The natives of Fizi thread the large *Turbo*. Among Solomn Islanders cowries are used as necklaces etc. In India cowries are worn as necklaces by children in village folk. The cowries also tied round the necks of cattles. Small bunches of *Terebellum* are worn as ear-rings. Toothed shells of *Dentalium* valued as ornaments.

5. Buttons and pearls

The shells of certain bivalves have been used for the mother of pearl layer used for buttons, knife-handles etc. Pearl buttons for clothing are cut from shells of freshwater bivalves but some marine clams are also used. The pearls obtained from clams and pearl oysters are among the most beautiful and valuable of jewels.

6. Ornamentation

Shell cameos are made mostly from snails, notable that of *Casis tuberosa* and *Cypraea tigris*. On one layer a design is carved and left unattached to an underlying layer of a different colour. The design may be white on a brown background or pink on a white background.

7. For Medicinal purposes

Preparations made from snails are valuable as a cure of many types of diseases and injuries. Five African slugs are roasted and powdered with half gram of *Acacia* and taken with myrth wine is an excellent remedy for desentry. Treated in various ways, the snails have been considered as cure for eye diseases, asthma, swelling of joints, headache, burns etc. Fossil cephelopods from Cretaceous strata, are collected by Red Indians of Montana and Wyoming as medicine.

8. In arts as dyes

The contents of the ink-sac of the cuttle-fish provide a rich brown pigment called "sepia" is used by artists. The original *indian-ink* was obtained from the ink of *Sepia.* Tyrian purple is obtained from juices of *Nucella* and *Murex.* The dye for royal purple came from the glands of *Murex.*

9. For illuminating purposes

In some parts of world the houses ae illuminated entirely by sky-lights windows, made up of shells probably the semitransparent valves of *Placuna.*

10. Religious purposes

The shell of chank (*Turbinella rapa*) is of special interest from its connection with the Hindu Religion. God Vishnu is represented by holding the shell in his hand, sinistral form of this animal is regarded with extra ordinary devotion. The chank is also blown at religious occasions.

HARMFUL MOLLUSCS

1. The clam that sinks ships

The ship-worm, *Teredo* has been the most destructive mollusc. It is able to scrape away wood by rotating the edges of shell back and forth against the end of the tunnel.

2. Destructive molluscs

Some gastropods are herbivorous. The land slugs and snails do considerable damage in gardens, greenhouses, mushroom beds etc. as they feed upon the succulent parts of seedlings and mature plants. Some gastropods are predators. *Nactica, Murex* etc. bore into and feed on other molluscs. The marine snail (*Urosalpinx*) causes serious losses to oyster industry. The snail is known as *oyster drill*, as it bores through the shells of oysters and clam to feed their soft bodies. Some gastropods are parasite on other molluscs. *Brachystomia* sucks blood from bivalve molluscs, polychaetes and echinoderms. *Octopus* feeds upon bivalves.

3. Shell with fatal stings

There are some cone shells whose sting is equally as powerful as the bite of rattle-snakes. The venom injecting apparatus lies in the head of animal.

4. Cradles and death traps

The giant clam (*Tridacna gigas*) measures about 1.2 meters in length. There are instances of tragic deaths of pearl divers who in advertently step into an open clam lying buried in the sand.

Index

P

R

S

T